The Role of FFR and Signal-to-Noise Ratio in Pitch Perception Research

Sita

Table of Contents

Chapter 1: Introduction 11

Background 11

Clinical Implications of this Proposed Study... 14

Specific Aims and Hypothesis 15

Chapter 2: Literature Review 20

Pitch Processing Overview 20

Factors that Affect Pitch Processing 22

The History of the FFR 23

Clinical Applications of the FFR 27

Common Dilemmas in FFR Recordings 29

Machine Learning 31

Non-negative Matrix Factorization 32

Subcortical Test Automation 36

Preliminary Studies 43

Study 1: Machine learning in Detecting FFRs in Chinese Adults: 44

Study 2: Machine learning in Detecting FFRs in American Neonates: ... 47

Study 3: Feasibility of Noise Separation in FFRs by using the NMF Algorithm. 50

Chapter 3: Methodology 54

Participants... 54

Recording and Stimulation Parameters.. 54

Data Analysis 55

Extraction of FFR Features 55

Non-negative Matrix Factorization 57

Statistical Analysis.. 58

Aim 1 58

Aim 2 60

Aim 3 60

Aim 4 61

Chapter 4: Results 63

Noise Separation by Using the NMF Algorithm 63

Application of the NMF Algorithm in Adults (Aim 1)... 68

Linear Regression 72

Exponential Modeling... 77

Two-way ANOVA.. 84

Evaluation of the NMF Performance for Adults (Aim 2).. 88

Linear Regression 91

Exponential Modeling... 94

One-way ANOVA 97

Application of the NMF Algorithm for Newborns (Aim 3a) 99

Linear Regression 102

Exponential Modeling... 106

Two-way ANOVA.. 112

Comparison of the NMF Performance Between Adults and Newborns (Aim 3b) ... 116

Linear Regression 119

Exponential Modeling... 123

Three-way ANOVA.. 128

Evaluation of the NMF Performance for Newborns (Aim 4a) 138

Linear Regression 142

Exponential Modeling... 145

One-way ANOVA 148

Comparison of the Hidden Response Between Adults and Newborns (Aim 4b) 150

Linear Regression 153

Exponential Modeling... 157

Two-way ANOVA.. 162

Chapter 5: Discussion 166

Understanding the Hidden Component of an FFR 166

Differences Between the ORI and NMF Conditions 170

Similarities and Differences Between the Adult and Newborn Groups 174

Optimal Parameters for Utilizing the NMF Algorithm in FFR Testing 176

Utilizing Machine Learning to Enhance Evoked Potentials 178

Study Limitations, Considerations, and Future Directions 181

Clinical Application 183

Chapter 6: Conclusions 185

References 186

Appendix A 198

Adult NMF Separation 198

Appendix B 201

Newborn NMF Separation 201

Appendix C 204

1 Iteration nSweeps Confusion Matrices 204

Appendix D 222

5 Iteration nSweeps Confusion Matrices 222

Appendix E 240

10 Iteration nSweeps Confusion Matrices 240

Appendix F 258

100 Iteration nSweeps Confusion Matrices 258

Appendix G 276

Exponential Modeling Tables 276

Chapter 1: Introduction

Background

Voice pitch plays an integral role in language processing (Oxenham, 2012). Properties of voice pitch have been the center of many notable studies (Anderson & Kraus, 2010; Carcagno & Plack, 2011; Jeng et al., 2011; Krishnan & Gandour, 2014). A common modality for tracking voice pitch is the Frequency Following Response (FFR).

The FFR is a type of electroencephalographical (EEG) recording that can be obtained by placing recording electrodes on the participant's head (Skoe & Kraus, 2010). FFR recordings display as a phase locked response to the envelope of a stimulus waveform. This method is shown to provide insight into the pitch processing abilities of the test participant (Anderson & Kraus, 2010). However, due to the inherently small response magnitudes in the FFR, the targeted neural responses are vulnerable to the magnitude of noise.

Noise in a recording can be defined as extraneous artifacts and any non-targeted neurophysiological activities that are asynchronous to the presented stimulus. In EEG recordings (including FFR recordings), background noise can adversely affect recording quality and therefore diminish the possibility of identifying a targeted response. The Signal-to-Noise Ratio (SNR) is a measurement between the targeted response in a recording (i.e., signal) and everything else picked up in the EEG signal (i.e., noise). Generally speaking, noise can be present as two sources in a recording. The first is the general background noise generating from outside the brain. The second is the natural noise that originates from inside the brain. Regardless of the different sources of noises, these noises impede the visibility of a targeted response. Such impedance presents a major challenge for researchers and clinicians who are interested in identifying targeted neural responses from potentially noisy EEG recordings. This issue is particularly apparent and challenging for difficult-to-test patients who are unable to respond behaviorally, such as young children, infants, and neonates.

In order to mitigate the adverse effects of noise and to improve the SNR of a recording, a common approach is to repeat a large number of recordings. This approach is titled signal averaging, and when implemented in a recording the method can aid researchers in understanding and enhancing the recorded EEG. In signal averaging, the recorded neural responses can be divided into two potions. The first is the "apparent" portion of the neural responses that stands out from noise through increased recording repetitions and second is the "hidden" portion of the neural responses that is not detectable through conventional averaging procedures. In traditional testing the first portion is what researcher recognize as the outputted EEG, while the second portion is encompassed in the recording but remains masked by the large magnitudes of noise. Ideally speaking the "hidden" portion could be apparent throughout the recording if there was a way to tease out the noise.

This traditional averaging method results in an increase in recording time, making EEG testing time extensive. In other words, a better SNR is required for detecting targeted responses, but takes a long time to complete. On the other hand, a shorter recording time saves time, but results in a poor SNR and thus is unable to detect targeted responses. Such a tradeoff between SNR and recording presents a major obstacle that prevents research from advancing in the realm of EEGs. The tradeoff between SNR and recording time has led researchers to seek out new innovations.

Non-negative Matrix Factorization (NMF) is a subset of machine learning which can be utilized for processing large-scale data through a matrix factorization model. When compared with traditional algorithms, this method has shown ease of implementation, good interpretability of the results, and a small storage space (Yu et al., 2014). NMF has become one of the popular multidimensional data processing tools in signal processing, biomedical engineering, pattern recognition, computer vision, and image engineering (Lin et al., 2015; 2017; Yu et al., 2014). Through the use of NMF algorithms, separations of noise from the response in an FFR recording can be achieved (Jeng, 2020, Unpublished data). This algorithm works by interpreting noise variants in each recording and extracting a similar pattern throughout a certain number of recording sweeps. The algorithm output consists of two components, the first being a gathering of all sorts of noises and the second a noise-free response.

A thorough investigation of the NMF algorithm must be conducted to identify the optimal parameters required to evoke an identifiable FFR. Once the optimal parameters are determined, these findings can then be applied to both reveal the hidden portion of an FFR recording as well as shorten recording times. Through the use of the NMF algorithm, clinicians and researchers can utilize this algorithm with maximal efficiency for screening, monitoring, and identification of individuals who may be at risk for certain disorders associated with decreased pitch processing abilities.

In order to bring utility to FFR testing, foundational research must be done to investigate how much of the original response is hidden by the magnitude of noise. Once this discovery is made, with the effects of noise removed, FFR testing time can be reduced. A significant reduction in testing time will allow the FFR to be implemented in day-to-day clinical testing. This will lead clinicians to highlight individuals who are at

risk for disorders associated with decreased pitch processing, such as kids with autism,

dyslexia, specific language impairment, and auditory processing disorder. Thus, the

overall goal of this study was to evaluate the performance of noise separation from FFRs

by implementing the NMF algorithm. With the hopes of contributing foundational

research, results of this study could drive forward potential clinical applications.

Clinical Implications of this Proposed Study

To this day, the issue of noise remains to be a perplexing factor in the estimation

of a response. The results of the feasibility study show that the NMF algorithm can work

on FFR recordings, but still requires more in-depth investigations. With a systematic

evaluation of NMF procedures, as well as uncovering the "hidden" portion of a response,

researchers can begin to understand in detail the applications this algorithm can provide

for FFR research.

The literature verifies that the FFR can display the potential for clinical

applications. It has been shown that when a stimulus is repeated real-time, sharpening

occurs in the brainstem leading to keen tuning and representation of speech cues related

to voice pitch. Since voice pitch can aid as an indicator in speaker identification, working

via following and separating a speaker's voice from other sound stimuli, a tool for

measurement and quantification of the system in imperative. As previously mentioned,

individuals with autism, specific language impairment, dyslexia, and auditory processing

disorders have displayed disruptions in pitch processing which has been expressed

through diminished FFR recordings. Unfortunately, the FFR is not utilized as a diagnostic

tool for the identification of these disorders due to the long recording time needed for testing. In clinical settings, testing must be time efficient and warranted by the clinician as necessary. Until either the time it takes to record an FFR is reduced or the noise in a FFR recording can be minimized efficiently, clinical utility will be just out of reach.

This study was conducted to help implement a new evaluation and identification method for patients who may face disorders associated with decreased pitch processing. The NMF algorithm shows prominent utility in both reducing the testing time of a recording as well as the noise in a recorded response. As stated previously, through comparative investigation of neonatal and adult FFRs, we show that the NMF algorithm aids in a reduction of recording time to bring the FFR into clinical testing scenarios.

Specific Aims and Hypothesis

Aim 1. To systematically manipulate NMF parameters and investigate their performances of noise separation from FFRs in adult participants.

In the NMF algorithm, three parameters were manipulated for the separation of noise from FFR recordings. These parameters are the maximum number of sweeps utilized in a recording (maxSweeps), the number of sweeps included in each subaverage (nSweeps), and the number of iterations (nIterations) implemented in the NMF algorithm. Three parameters were systematically manipulated to see which parameter combinations would produce optimal results. In sum, the goal of this first aim was to establish optimal parameters required to obtain an identifiable FFR, with the hopes of reducing test time.

Hypothesis. It was hypothesized that the performance of the NMF algorithm, reflected by the measurements of FFR features, would increase and reach a plateau with increasing maxSweeps, nSweeps, and nIterations. Specifically, based on the work by Skoe and Kraus (2010) that an FFR can be obtained by including 1000-6000 sweeps, it was hypothesized that an increment of the FFR features should be observed with increasing maxSweeps and would reach a plateau occurring approximately at 1000 maxSweeps. Additionally, based off the work reported in Jeng and colleagues (unpublished data) that utilized similar protocols in noise separation through NMF algorithms, FFR features would increase with increasing nSweeps and would reach a plateau occurring at around 1000 nSweeps. When nIterations are manipulated, an increase in FFR features will be visualized with increasing nIterations and a plateau should be observed around 100 iterations, as reported in Hart and Jeng (under review).

Aim 2. In this study, we present a novel approach to extract the "hidden" portion of neural responses, as well as the "apparent" portion of neural responses.

Signal averaging is a process which utilizes a large number of recording repetitions to aid in improved SNR. This technique has been used to extract neural responses from EEG recordings for decades (Hyde, 1994). Researchers have shown that, when a certain number of sweeps are included in a recording, the amplitude of noise is reduced. This results in enhancing the visibility and detectability of the neural responses. Another way to explain this method is, when a certain number of sweeps are included, portions of the neural responses "stand out" from surrounding noise and become visible and detectable to the experimenter. Similarly, there are potions of the neural response that are "hidden" or "masked" in the recording due to the dominating magnitude of the noise. Based on the understanding of signal averaging, neural responses can be divided into two potions: (1) the "apparent" portion of the neural responses that stands out from noise through conventional averaging procedures and (2) the "hidden" portion of the neural responses that is not detectable through conventional averaging procedures.

Therefore, through the use of the NMF algorithm, a thorough investigation of the "hidden" response can be implemented. From this procedure measurements can be made to determine if the NMF algorithm can be used to overcome the technical pitfalls that researchers face when using conventional averaging methods. Estimations based off the noise amplitude and the response amplitude can aid in the understanding of an EEGs response components.

Hypothesis 2. It is hypothesized that the "hidden" portion of neural responses in adults will decrease with increasing number of sweeps. As the noise amplitude declines with increasing sweeps, a larger portion of the responses will become more apparent from the portion of neural responses that were originally hidden at a fewer number of sweeps. When the NMF algorithm is applied it is hypothesized that the "hidden" portion of the neural response will be apparent (i.e., larger in magnitude) at a lower sweep count than that at a higher sweep count when utilizing traditional signal averaging.

Aim 3. To systematically manipulate NMF parameters and investigate their performances of noise separation from FFRs in neonates.

NMF procedures may be applied to data from distinct age groups in order to understand the mechanisms behind pitch processing across ages. Given the differences between neonate and adult nervous systems, a systematic investigation of NMF procedures should be applied and compared. Neonatal FFRs have been studied by researchers over the past few years and have exhibited recordable responses in a majority of neonates during their immediate postnatal days. While neonatal FFRs have shown smaller responses compared to adult recordings, they are identifiable when certain precautions are followed. Although the neural circuitries at the subcortical level are still developing over the first few days after birth, the robustness of a response can still be identified. Based on the ability to record identifiable responses during immediate postnatal days, neonatal FFRs could potentially be used to evaluate pitch processing. In children, FFR findings have shown a diagnostic potential in highlighting disorders associated with pitch processing, such as autism (Russo et al., 2008), specific language impairment (Rocha-Muniz et al., 2012), dyslexia (Chanderasekaran et al., 2009; Ramus, 2012), and auditory processing disorder (Rocha-Muniz et al., 2012). Despite the importance of recording FFRs in neonates, the FFR by nature is a small-amplitude response, making it susceptible to noise. Neonates are impulsive with their physiological movements; this impulsivity increases noise in FFR recordings. This variability makes testing neonates more difficult and time consuming when compared to adults. The balance between recording quality and the time it takes to obtain the response is important. As mentioned, the FFR is a small-amplitude response meaning more recording repetitions are warranted to separate the response from the noise in the recording.

Hypothesis. It is hypothesized that the performance of NMF algorithms, reflected in the measurements of FFR features, in neonatal recordings will reveal similar results compared to the anticipated outcomes of adult recordings. Specifically, it is anticipated that neonatal FFRs will require a higher number of maxSweeps, nSweeps, and nIterations to accomplish comparable performance to adult recordings. For example, the slope of the FFR features will be shallower than that of the adult recordings due to the noisy nature of neonatal recordings (Jeng et al., 2018). The plateau for neonatal data will occur at a higher number of maxSweeps, nSweeps, and nIterations than the adult recording given neonatal recordings are much noisier than adult recordings (Jeng et al., 2018).

Aim 4. To evaluate the "apparent" and "hidden" components of an FFR response in neonates.

As mentioned in Aim 3, it is important to understand the mechanism behind pitch processing across ages. When testing neonates, recording sessions are traditionally longer than those of adults. This time increase is due to the inherently larger variability in noise when testing neonates. Based off of a conventional signal averaging, with neonatal recordings, more sweep accumulations are required to tease out the "apparent" portion of the recording from the "hidden" portion. If an algorithmic approach could be implemented to reveal such portions in a more time efficient manner, the FFR could be implemented in clinical testing of neonates as young as 1-3 days old. Early testing and monitoring could be crucial in the identification of individuals who may face disorders associated with decreased pitch processing, such as autism, dyslexia, auditory processing disorder, and specific language impairment. Through the use of NMF procedures on neonatal recordings, the development of FFR automation may begin. Findings may lead to clinical implementation of the test for individuals during immediate postnatal days. Hypothesis. Similarly, to hypothesis 2, it is hypothesized that the "hidden" portion of neural responses in neonates will decrease with increasing number of sweeps. As the noise amplitude declines with increasing sweeps, a larger portion of the responses will become more apparent from the portion of neural responses that were originally hidden at a fewer number of sweeps. When the NMF algorithm is applied it is hypothesized that the "hidden" portion of the neural response will be apparent at a lower sweep count than that of the traditional EEG utilizing signal averaging.

Chapter 2: Literature Review

Pitch Processing Overview

The way individuals make sense of incoming sound stimuli is imperative for daily communication. Pitch is a conversational characteristic that is commonly defined as the highness or lowness of a tone perceived by the listener. In non-tonal languages, pitch fluctuations can dictate the meaning of a statement. Ending a sentence in a higher pitch may lead someone to think that a question is being asked, whereas ending in a flat or lower pitch can indicate a statement. In tonal languages, a change in voice pitch can alter the meaning of the word being said. The importance of pitch is shown throughout numerous studies. In order for individuals to understand conversations with varying pitch, they must have adequate pitch processing. From this importance, pitch processing has become a studied topic over the years giving insight to various communication disorders.

Early studies have shown that FFR reflects spectro-temporal details of the stimulus with high fidelity (Skoe & Kraus, 2010; Wong et al., 2007). Given that the FFR faithfully reflects spectro-temporal detail, such a response has provided key insights into information related to auditory pitch processing. Pitch processing has been shown to develop throughout an individual's lifespan. At 30 weeks of gestational age, a fetus possesses the capabilities to hear sound stimuli (Listovsky, 2015). When an infant is born, they possess the capabilities to detect a majority of phonetic variations which can be found in human languages (Eimas & Miller, 1992; Kuhl, 1979; Kuhl et al., 2006). This notion supports the biological capacity model, stating that speech perception is influenced by inherent factors. This model states that, when individuals are born, they can already process speech and pitch cues and possess the capabilities to interpret and understand such nuances throughout development. Conversely, the linguistic experience model specifies that infants' processing abilities are developed through linguistic exposures within their acoustic environments (Kuhl et al.,1992). Throughout an individual's development, sound is constantly shaping and molding speech interpretation and pitch processing (Krishnan & Gandour, 2014).

Although it has been shown that speech perception in neonates and infants display similar patterns, research has shown evidence supporting that speech perception is more heavily related to innate factors than language experience (Kuhl, 2010). It is important to note that through enriched experiences, such as language or musical exposures, can also enhance pitch processing in individuals (Krishnan and Gandour, 2014).

As the brain matures to its adult form, the FFR becomes better defined. Due to the increase in myelinated neurons in the nervous system, neural conductions occur at a faster rate and more accurately, resulting in better FFR recordings for adults than infants. Another reason that adult FFRs are more defined is due to the fact that neonates are more restless than adults. Adults have a better control of autonomic functions than neonates. As a result, when recording FFRs from adults, a response can become identifiable at a lower number of sweeps. The need for a larger number of sweeps to be included in an averaged waveform is needed less for adults when compared to neonates. In neonatal recordings, a higher number of sweeps is required to separate the response from the background noise. The fundamental differences between adults and neonates have been shown in several studies (Gardi et al., 1979; Jeng et al., 2018). The differences and similarities between these two age groups highlight where research needs to be conducted to benefit future utilization of the FFR.

Factors that Affect Pitch Processing

Various factors such as language experience, hearing sensitivity, and cognitive function can impact pitch processing abilities. As humans age, they are influenced by the acoustic environments around them. In a previous research study, Krishnan and Gandour (2014) discuss the effects of language experience in the human auditory system. They stated that through the enrichment of speaking and listening to a tonal language, native speakers showed stronger neural representation than nonnative speakers. Such experience-dependent responses support the enhanced development of pitch processing in individuals with tonal language exposure.

This dependency also ties into hearing sensitivity as an important factor on pitch processing. Sensory inputs such as hearing sensitivity and audibility are important to enhance and develop the auditory system. Without such influences, the auditory system may not develop properly (Listovsky, 2015). In a study conducted by Oxenham (2008), individuals with hearing loss were compared to normal hearing listeners based on their perception of pitch. Oxenham (2008) found that individuals with hearing loss had less access to individual harmonics when compared to their typical hearing peers. Such coding deficits were thought to be the cause of many difficulties expressed by individuals with hearing loss such as, difficulty with speech in noise, difficulty attending to a conversation, and difficulty distinguishing more than one pitch at a time.

Research in individuals with hearing loss has shown that, with a sensory disconnect, auditory deprivation can occur and lead to cognitive decline or brain degradation (Cherko et al, 2016; Lin et al., 2014). For example, studies have investigated the effects of individuals with language and learning disabilities. Individuals diagnosed with autism, auditory processing disorder, specific language impairment, and dyslexia exhibit general limitations in encoding complex speech information (Banai et al., 2007). In regard to pitch processing, individuals with these diagnoses also exhibit degraded abilities in pitch processing when compared to their matched peers (Chanderasekaran et al., 2009; Ramus, 2012; Rocha-Muniz et al., 2012; Russo et al., 2008). In summary, exposure to specific languages, hearing loss, and cognitive deficits have been shown to affect an individual's language processing centers and share similar methods in quantifying pitch deficits.

The History of the FFR

Originating in the early 1970s, the FFR is a measure of phase-locked neural activities, which represent the brain's ability to encode pitch characteristics (Moushegian et al., 1973). Preliminary FFR testing began with experimental studies to investigate neural generator and to localize the source in animals such as cats (Smith et al., 1975) and chinchillas (Hou and Liscomb, 1979). In 1979, Gardi and colleagues recorded FFRs from cats and concluded that multiple neural generators contributed to the overall spectro-temporal properties of the acoustic stimulus. Specifically, the cochlear nuclei contributed to approximately 50% of the FFR amplitude, the cochlea contributes roughly 25% (when only one polarity of the stimulus is provided), and the superior olivary nuclei accounts for about 20% of the amplitude. In contrary, they noted that the influence of the inferior colliculus was found to be insignificant. When toneburst stimuli are varied from 300-3000 Hz, FFR latencies were prolonged with increasing carrier frequencies. The authors concluded that, in addition to physical properties of the basilar membrane, neural generators at differing levels of the auditory pathway might also contribute to the prolonged latencies. Although the researchers did not investigate any changes in FFR for cats with hearing loss, their results suggested that identifiable FFRs could be recorded in humans if hearing loss and audibility are compensated.

Beginning in the early 2000s, more research in human FFRs began to spark. In 2000, Ballachanda and Moushegian looked into the effects of interaural time and intensity differences in normal hearing adults. Major conclusions from their study revealed that the FFR possesses the potential to become a tool for the identification of normal and abnormal processing along the auditory pathways. In 2001, Galbraith and colleagues conducted a study to investigate the effects of recording montages in FFRs. They found no significant latency difference between the multiple horizontal montages, but the vertical montage resulted in prolonged latencies. Results from these findings showed clinical relevance and adaptive potential in human studies, therefore driving future research endeavors in the field.

As FFR research began to unfold, the stimulus that could be used to elicit a response also adapted and evolved. Early FFR studies began with the use of simple pure tones (Worden & Marsh, 1968; Marsh et al., 1970). Throughout various trials, it was found that complex stimuli, such as speech, could also be used to elicit an FFR (Galbraith, 1994; Galbraith et al., 1995). Furthermore, a wide spectrum of varying stimuli have been utilized to elicit FFR responses such as tone bursts (Glaser et al., 1976; Moushegian et al., 1973), two-tone stimuli (Gokel et al., 2012; Rickman et al., 1991; Smalt et al., 2012), multi-tone complex (Gardi et al., 1979; Zhang and Gong, 2019), synthesized speech signals (Krishnan, 2002), two-tone approximation of vowels (Plyler & Ananthanarayan, 2001), and natural human voices (Banai et al., 2009; Hornickel et al., 2009; Jeng et al., 2010; Johnson et al., 2007; Krishnan et al., 2004; Musacchia et al., 2007; Parbery-Clark et al., 2009).

After the basic aspects of the FFR have been studied from the late 1970s to the early 2000s, researchers are now beginning to incorporate the FFR to its full research and clinical potentials. The importance of FFR stems from the high fidelity in reflecting many of the important spectro-temporal details that are embedded in the stimulus. The FFR closely mirrors the morphology of the stimulus, therefore displaying the same pitch properties processed in the brain. Recorded responses from the listener can lead researchers to make inferences on their brains' abilities to preserve and follow the stimulus morphology. Most importantly, such inferences can also show the significance the FFR displays in how individuals process and make sense of incoming acoustic stimuli for daily communications (Chandrasekaran et al., 2009; Kraus & White-Schwoch, 2017; Tallal, 2004).

An advantage to using the FFR in research settings is related to its consistency (Song et al., 2011). Since it is a subcortical measure, the FFR does not require an individual's attention, cognition, or participation throughout testing. Therefore, the FFR is an ideal test for patients who cannot provide a response in traditional behavioral testing environments. On the contrary, in cortical test measurements, patients must actively participate in the task at hand. This becomes a difficult process when beginning to test individuals of younger or disordered populations. Hearing estimations, functions of the auditory system, and processing abilities are desired to be obtained in such individuals. A constriction is that traditional test methods can provide only a limited amount of information about pitch processing.

Clinical test batteries that assess hearing estimations, auditory functions, and pitch processing consist of behavioral methods which require active participation throughout the entire testing session. These methods are tedious and cannot always be completed by some patients. There are different test measures which can assess these similar domains without the need of patient participation. Standard objective test measures such as the Auditory Brainstem Response (ABR) and the Auditory Steady State Response (ASSR) can assess hearing thresholds and functions of the auditory pathways in younger populations and difficult to test patients. Despite the fact that the ABR and ASSR can provide information about hearing thresholds and auditory functions, these measures fail to highlight how the brain processes and encodes sound stimuli in detail. Both the ABR and ASSR in clinical applications traditionally give light to neuroanatomical function in regard to hearing thresholds, but not pitch processing abilities.

Therefore, researchers have investigated FFR testing throughout different stages of life to evaluate its clinical relevancy. For example, Jeng and colleagues (2010) sought out to assess subcortical processing in neonates 1-3 days old. Their study concluded that, at the age of as early as 1-3 days old, pitch processing capabilities were present and could be measured through an FFR. From these findings, more studies began to assess the functionalities of the neonatal auditory pitch-processing system (Anderson et al., 2015; Jeng et al., 2011; Jeng et al., 2016; Ribas-Prats, 2019). As neonates advanced in age, it was found that the recorded FFRs became better defined but were similar in amplitude when compared to the results recorded at a young age. These observations have served as a basis for research studies to this day and may also aid in the scientific backing of detecting pitch-processing disfunctions at such an early age.

Clinical Applications of the FFR

In the literature, there is a distinct difference between ABR and FFR recordings. For example, children with language-based learning problems exhibited decreased FFRs, despite normal encoding revealed via click evoked ABRs (Banai et al., 2005; Johnson et al., 2007; Russo et al., 2009; Song et al., 2006). Such a difference has led researchers to reevaluate the use of the ABR in clinical testing. Researchers postulated how individuals could have a normal ABR but expressed a diminished FFR. Although the ABR should be sensitive to functional disconnects in the auditory system, there are certain functionalities which cannot be assessed via ABRs but is assessable via FFRs. While the ABR is beneficial in evaluating subcortical structures functionality, it is lacking in providing descriptive data in terms of pitch processing.

As mentioned, the FFR is sensitive to disorders associated with decreased pitch processing. In 2008, Russo and colleagues conducted a study in which they observed a difference between FFR recordings of individuals with autism and typically developing (TD) peers. Findings concluded that children with autism exhibited decreased pitch tracking abilities when compared to their TD peers. Furthermore, researchers postulated that this non-specific, non-periodic, disconnect in their pitch processing could be related to these individual's brainstem not recognizing the fundamental frequency as an important acoustic clue. For investigations of individuals with auditory processing disorder and specific language impairment, Rocha-Muniz, Befi-Lopes, and Schochat (2012) discovered that children diagnosed with either disorder displayed abnormal neural encoding and temporal deficits. Notably onset responses were asynchronous and prolonged when compared to their TD peers. This finding is consistent with the results noted in the study conducted by Russo in 2008, relating back to these perceptual deficits, the nature of this delay could be attributed to structural underdevelopment of the brainstem.

In studies conducted by Ramus (2014) and Ramus and Ahissar (2012), the authors assessed FFRs recorded from children diagnosed with developmental dyslexia. Ramus (2014) explained that, although developmental dyslexia is considered to be attributed to higher order processes, there are some underlying mechanisms that remain unknown and need to be researched. These findings aid in the rationale for the FFR serving as an evaluator for individuals with developmental dyslexia. In a study conducted by Chandrasekaran and colleagues (2009), the researchers sought out to better understand the encoding patterns of individuals with developmental dyslexia. Findings concluded that children with developmental dyslexia displayed FFRs which could not process nor follow the spectro-temporal details of the test stimuli when compared to TD counterparts. In particular, researchers noted that individuals in the "poor readers" group showed a deficit in perceiving speech in noise, specifically they could not adapt to the repeating elements of the incoming signal. This deficit results in diminished function when individuals are listening to a targeted signal in background noise. Findings from this study conclude that the overall deficit in pitch signal extraction may contribute to the deficits exhibited in noise separation.

The FFR findings from the different disordered populations listed above can serve as clinical backing to support early identification and intervention. Given that FFR testing can be achieved in individuals as young as one to three days old (Anderson et al., 2015; Jeng et al., 2016; Jeng et al., 2011), the implementation of screening and monitoring methods could be achieved. The need for more research and data collection with FFRs is warranted to overcome any possible technical issues which may impede on testing ease and implementation.

Common Dilemmas in FFR Recordings

To obtain a quality FFR recording, a large number of recording sweeps are required. This stems from the tradeoff between recording sweeps and the Signal-to-Noise Ratio (SNR). Because the FFR is such a small amplitude response, a large number of recording sweeps should be included to minimize the noise in the recordings. This method serves to provide sufficient signal averaging. Averaging is based off the principle that noise in the recording is asynchronous to the stimulus (Hyde, 1994, Ruchkin, 1988). After a recording the EEG is segmented into time blocks called sweeps. Typically, each sweep has a time interval containing responses to the stimulus as well as noise. Because each recording sweep contains the response as well as the noise, averaging may be used to minimize the influence of noise. Within each recorded sweep, the noise likely will vary over time, and thus asynchronous to the stimulus. It should be noted that each sweep may also contain neural activities that are synchronous to the stimulus. In other words, averaging will work in a manner where anything that is the same and synchronized to the stimulus will remain in the averaged waveform and anything that varies and is not synchronized to the stimulus will be averaged out. With an increase in recording sweeps, averaging will become more effective. This will result in minimized noise and boost the overall SNR of the recording.

Averaging can be described by the equation:

$$SNR = \frac{response\ amplitude}{noise\ amplitude} \times (Number\ of\ sweeps)^{1/2} \qquad (1)$$

In summary, with increasing numbers of sweeps, the noise can be averaged out and thus improving the SNR of a recording.

Typically, the number of sweeps used for an FFR recording may range from 1000 to 6000 sweeps (Skoe & Kraus, 2010), or even up to 8000 sweeps (Jeng et al., 2018). This need for a large number of sweeps lengthens the recording time. This can become problematic when testing participants who may not be able to remain still or asleep throughout the entire recording session. To provide an example of the time it takes to complete an FFR recording, with the use of the American vowel /i/ with a rising frequency contour that is commonly used in our lab, it takes approximately 39.3 minutes to obtain 8000 accepted sweeps (i.e., (250 ms + 45 ms silent interval) x 8000 sweeps / 1000 = 2360 seconds / 60 = 39.3 minutes). When considering the time that is required for skin preparations and electrode attachments, the length of a recording session can span up to one hour or longer even in the most ideal situations. In addition to the time required for skin preparations and electrode attachments, high levels of artifacts may also increase the recording time substantially. Knowing those limitations, it is typical that only one recording can be obtained per participant. This limits the amount of data available for subsequent analysis in FFR studies. Given the clear tradeoff between a quality recording and the time it takes to obtain an FFR, researchers have to seek new innovations to overcome such problems.

Machine Learning

The use of machine learning has become more prevalent in today's technologic advancements. Machine learning assists computer systems with enhancements of pattern recognitions and feature extraction to analyze data without the need for human judgement or supervision (Marsland, 2015). Machine learning is ingrained into the technology used in day-to-day activities for all individuals. For example, through the use of facial recognition or fingerprint unlocking of cell phones and personalized medical treatment, various industries are tapping into machine learning in order to minimize human error and ease day to day tasks.

Previous researchers have investigated the effects of implementing machine learning in the identification of soundscapes. A soundscape is described as varying sound components existing simultaneously in an environment. Within each soundscape, there are multiple components that can be extracted for data analysis. In a paper by Lin and colleagues (2017), soundscape recordings from various bat caves in national forests were collected. Post data collection, the researchers utilized NMF procedures to tease out various noise components (i.e., road noise, wind noise, other animal sound) that are unrelated to the targeted sound source (i.e., bat calls). In the literature to date, recordings processed through NMF procedures have been used predominately in examining animal diversities, environmental factors, and their relationships. For example, these procedures have been applied successfully to the vocalization, health, and diversities of birds, bats, fish and whales (Lin et al., 2015; Wall et al., 2013; Walters et al., 2013; Wimmer, 2015).

There is a similarity between soundscape experiments and FFR recordings. Soundscape recordings consist of a targeted sound source as well as background noise.

This is similar to the neural response and noise components in FFR recordings. Given the success of utilizing the NMF algorithm in separating background noise from a targeted sound source, FFR response components may also be extracted in a similar manner. In other words, researchers could separate physiological background noises from a targeted neural response. If successful, the results would contain two components. The first component is the physiological background noise and the second component is the targeted sound source. In sum, the NMF algorithm could be utilized with FFR recordings to separate noise and reveal the component that is predominately the targeted neural response. With the use of machine learning, researchers can now work toward creating an automated system that will enable faster response recognition of FFR recordings.

Non-negative Matrix Factorization

NMF is a subset of machine learning which can be implemented for signal processing and feature extraction. NMF is particularly suitable for tasks where input factors can be interpreted as non-negative values. This method has been shown to be beneficial in the decomposition of multivariate data (Lee & Seung, 1999), and thus should be feasible in separating noise from FFRs.

NMF may also be considered as a special type of machine learning, falling under the unsupervised learning category, based on its clustering patterns. NMF works through variations of the equation:

$$V = W \times H \tag{2}$$

Where V is the original input matrix, W represents the weighting factor, and H reflects the targeted sound source. In reference to FFR recordings, V will be the recorded EEGs, W will represent the noise in the recordings, and H will reflect the targeted neural response. To display the general workflow of the NMF algorithm, Figure 1 depicts the basic procedural steps and optimization cycles that are involved to fine tune the algorithm performance. From this process, many studies have shown the benefit of the NMF algorithm in the identification of noise, facial images, protein evolution, and more (Lee & Sung, 1999; Lin et al., 2017; Yu et al., 2014). Despite subtle variations in the computations of NMF algorithms, the general purpose of the algorithm remains clear in its methods for signal processing and feature extraction.

There are three main advantages of NMF algorithms (Yu et al., 2014). First, the algorithm enforces a non-negative constraint. In other words, the input matrix cannot contain any negative values. Furthermore, NMF ensures that negative values cannot appear in the decomposition results of W and H, due to the use of the multiplicative update rule. This non-negative constraint proves useful in many applications that utilize the NMF algorithm. Through the use of an internal constraint (i.e., the non-negative restriction), the capacity of the algorithm proves to function more efficiently than a traditional external restriction of non-negativity. This means that many other algorithms utilize an external constriction, indicating no negative values can be present in the input matrix. Conversely, with the NMF algorithm, such a concern is eliminated through the use of the internal non-negative constraint. This advantage of the NMF algorithm should be applicable and beneficial with FFR recordings, given there should be no negative values included in the analysis. For example, when depicting the amplitudes on a spectrogram (z axis) the nature of the sound amplitude should indicate no negative values.

The second advantage is that NMF provides information about the relationship between two input variables without the need for complex analysis techniques. For example, when Principal Component Analysis (PCA) is used, orthogonality is assumed among input variables. On the other hand, when the NMF algorithm is used, the W and H resultant matrices do not need to be orthogonal to each other. Therefore, NMF overcomes a traditional problem of a PCA. In a PCA, there may be issues with displaying complex data due to orthogonal constraints. Simply put, the relationship displayed in a PCA may not reflect the true nature among the input variables. This issue is apparent when the input variables are not orthogonal to each other. On the other hand, the NMF algorithm is free of the orthogonality constraint. Therefore, the NMF results may better reflect the relationships among input variables. Overall, the NMF algorithm appears to provide a better representation of the data when compared to standard complex techniques, such as the PCA. Another triumph is that, since the input variables in an NMF cannot be negative, the issue of negative elements in the PCA is no longer a concern.

The third advantage of the NMF algorithm is the ease of implementation. For a PCA to estimate a targeted sound source, multiple recordings should be obtained from different locations. For the NMF algorithm to identify a targeted sound source, multiple recordings from a single location will suffice. Another benefit is that the NMF algorithm provides a two-dimensional output (e.g., H), whereas the PCA provides an estimated neural component as a vector. Although both algorithms are considered point-to-point calculations, it is possible that a two-dimensional output matrix will contain more enriched details than a simple vector. Furthermore, the NMF algorithm may provide a method which analyzes data more thoroughly than standard complex techniques, such as the PCA.

In summary, the three main advantages of NMF implementation depict the positive utilization of the algorithm in soundscape separations. When applied to FFR recordings, the NMF algorithm should be able to provide appropriate separation between the noise and the targeted neural response. If successful, this method could potentially aid in the development of test automation and bring FFR measures to clinical light.

Figure 1

A flow chart of the NMF algorithm

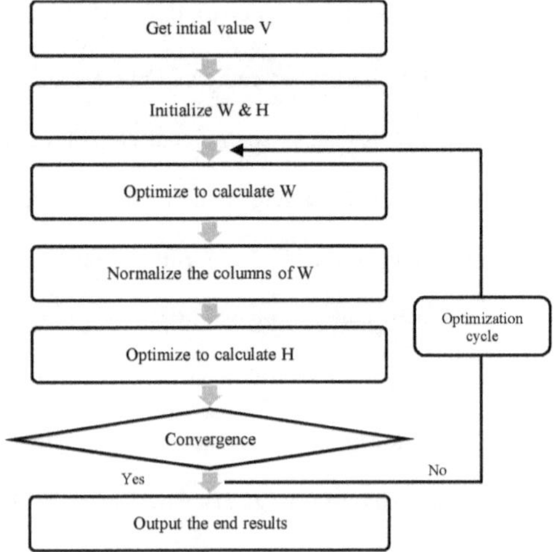

Subcortical Test Automation

Some tests in the hearing science field are already being utilized in automated procedures, such as the ABR and ASSR. These tests are implemented to assess hearing functioning through a pre-defined stop criterion which enables the termination of a recording when the criterion has been met (Delgado & Ozdamar, 1994; Valdes et. al., 1997). The creation of automated ABRs began when researchers (Don et al., 1984; Elberling & Don, 1984; Özdamar et al., 1994; Sininger, 1993) focused on creating statistical algorithms to determine the presence of an ABR. As a result of these research endeavors, the ABR that was once strictly a research tool had become a clinical estimator of hearing sensitivity and neural dysfunction.

As mentioned preciously, the ABR was originally used as a research tool which aided in the evaluation and assessment in auditory neural generators along the auditory pathway. Since the ABR is a small-scale evoked response, it is subject to noise interference. This downfall restricted the ABR's clinical potential and withheld the test from being applied in many test settings. Slowly, the ABR was introduced into clinical settings. Throughout its development, the lack of automation was found to be a major obstacle limiting the ABR's clinical potentials. In the late 1970s-early 1980s, researchers made foundational efforts to create an automated ABR system, but manual interpretations by the operator were still needed (Schimmel et al., 1974; 1975; Wong & Bickford, 1980). These findings led researchers, such as Eberling and Don (1984), to begin the detailed works in ABR test automation. Their preliminary study began with a statistical approach for detecting the presence of an ABR by calculating an F_{sp} ratio between the background noise and the estimated variance of the ABR waveform. The results of this study laid the theoretical and mathematical baseline for future studies in ABR automation.

As a follow up to the research provided in the first study, Don, Eberling, and Waring (1984) then applied the procedures to demonstrate their capabilities in automatic threshold detection. The process works by calculating an F_{sp} value, which can be obtained through the following equation:

$$Fsp = \frac{VAR(S)}{VAR(SP)} \tag{3}$$

Where F_{sp} is the F-statistical value referenced to a single point, VAR(S) is the variance of the averaged waveform, and VAR(SP) is the variance of the background noise at a single point among individual sweeps. The F_{sp} value, when computed, represents a result closely related to the SNR of a recording. Because the F_{sp} value is computed based on statistical procedures, it allows the examiner to determine the level of confidence whether an averaged waveform contains a true response or not. That is, an F_{sp} value can be calculated for every 256 sweeps to help provide a determination of the presence or absence of an ABR. For example, if the F_{sp} value exceeds the 95% confidence intervals of the data variances, a response is considered present. On the other hand, if the Fsp value is within the 95% confidence intervals of the data variances, a response is considered absent.

The F_{sp} computation is based on three factors: response amplitude, noise amplitude, and the number of sweeps. In other words, there are three methods that can be used to enhance the F_{sp} value. The first method is to try to increase the response amplitude. Unfortunately, the amplitude of a response is usually within a fixed magnitude range, when a specific stimulus is presented at a certain intensity to the listener. The second method is to attempt to reduce the amplitude of the noise. The best way to reduce noise amplitude is to eliminate the noise sources such as, turning off unnecessary electronic devices in sound booth, checking and maintaining adequate impedances, and providing proper head and neck supports for the patient. The third method is to increase the number of sweeps included in the averaged waveform. In other words, through the inclusion of a larger number of sweeps, the F_{sp} value will be enhanced. Arguably, the greater number of sweeps included in the averaged waveform will increase computational demands that are required to provide an F_{sp} value. However, with the power of modern technologies, those computational demands are rarely an issue nowadays.

In addition to the three factors that can influence the F_{sp} value, other factors such as participant restfulness and noisy recording environments can adversely affect the F_{sp} value. For example, if a participant is unable to relax during a recording, it will result in a higher noise amplitude when compared to recordings that are obtained from participants that could relax. In this case, a larger number of sweeps may be required to provide an adequate F_{sp} value. In other words, the quality of a recording is inversely related to the number of sweeps required to obtain a statistically significant F_{sp} value. If there is a poor-quality recording, more sweeps will be required to distinguish a response from background noise. Conversely, if there is a good quality recording, fewer sweeps are required.

It should be noted that the F_{sp} computation does not require any operator interpretation of the recordings. The need for operator assessment of the response is eliminated when using this computation. Based upon these calculations, ABR recordings can be evaluated and assessed without operator influence (Sininger, 1993). Furthermore, with attempts to apply F_{sp} in clinical settings, various companies have adapted, modified, and utilized this algorithmic procedure to create automated ABR systems that are commercially available.

The ALGO system, developed by Natus, is another method that provides ABR automation. To date, the ALGO system is one of the most commonly utilized in ABR automations for neonatal and infant recordings. Specifically, the ALGO system utilizes a correlation between the averaged waveform that is just obtained from a participant and an ABR template waveform that is obtained from a number of normal hearing neonates and infants that are in the same age group as the participant. In other words, the higher the correlation value, the higher likelihood it is for a present response. As such, the correlation value is a statistical product indicating the likelihood of a presence of a response. It should also be noted that such a statistical procedure does not require human judgements and thus is a legitimate method for ABR automation.

The ALGO system works to aid in the identification of ABRs from neonates and infants when acoustic clicks are presented at 35 dB nHL. Through research, it is recommended to use a click rate of 37/sec, with an attempt to shorten recording time. For example, it would take roughly 45 seconds to obtain 2000 recording sweeps. This approach could provide a short recording time window, therefore enabling a feasible recording timeframe. With such a short window to complete a recording, the ALGO system could be used in clinical settings (Jacobson et al., 1990). Within the ALGO system, the correlation value can be updated when every additional 500 sweeps are added to the averaged waveform. Once the correlation value exceeds the stopping criteria, the recording can be terminated due to the likelihood of a response being high. In addition to the ALGO system developed by Natus, other companies such as Bio-logic, Vivosonic, Interacoustics, and Maico have also developed similar automated ABR systems.

Similar to the automated procedures developed for the ABR, researchers have also developed automated procedures for the ASSR (John et al., 1998; John & Picton, 2000; Vander Werff, 2009). The ASSR is an electrical potential evoked by presenting sinusoidally amplitude-modulated tones at specific carrier frequencies. A unique feature of the ASSR is that its response energy is located at the modulation frequency, whereas the energy of the stimulus is located at the carrier frequency and its two side bands (John & Picton, 2000). One way to conduct ASSR automation is to determine the relationship between the response amplitude at the modulation frequency and the spectral amplitudes of 512 frequency bins on both sides of the modulation frequency (i.e., 256 frequency bins above and below the modulation frequency) (John et al., 1998).

ASSR automation works by using an acoustic stimulus at a specific modulation frequency. The automation process typically begins by dividing incoming EEG data into sweeps. Within each sweep, there are 16 epochs. Within each epoch, it may contain eight or sixteen modulation cycles. As a matter of fact, any integer multiples of a modulation cycle may be included and defined as one epoch. A Fast Fourier Transform (FFT) can then be performed on the recorded sweeps to convert the recorded time waveforms from the time domain to the frequency domain. Once converted, an F-Ratio statistic can then be used to evaluate the energy at the modulation frequency to determine the presence of an ASSR. The FFT is repeated when more sweeps are included to the analysis. When the

FFT is performed on a cumulation of recorded sweeps, it may improve the F-Ratio at the modulation frequency. The calculation for the F-Ratio is:

$$\frac{(x_s^2 + y_s^2)/2}{(\sum_{i=1}^{j} x_i^2 + \sum_{i=1}^{j} y_i^2)/(2j)} \tag{4}$$

Where x_s is the sine term of the polar vector at the modulation frequency, y_s is the cosine term of the polar vector at the modulation frequency, x_i is the sine term of the polar vector at the frequency i, and y_i is the cosine term of the polar vector at the frequency i. Additionally, i and j are the lower and upper limits of the frequency range that is used to estimate the magnitudes of the background noise.

When the estimated F-Ratio reaches significance (generally $p < 0.05$) at the modulation frequency, an ASSR can be determined as present (John et al., 1998; John & Picton, 2000). Such an approach can be utilized for recordings that are evoked by using one single modulation frequency. Alternatively, such an approach can also be utilized for recordings that are evoked using multiple modulation frequencies in both ears.

One of the first automated ASSR systems was developed by John and Picton (2000) at the Rotman Research Institute at Toronto University. They titled it as the MASTER (Multiple Auditory Steady-State Response) system. The MASTER system works by focusing on multiple modulation frequencies that are delivered to the participant simultaneously. Based on the success of the MASTER system, different versions of ASSR automations were also developed and implemented commercially. Companies such as the Bio-logic Navigator Pro EP system by Natus, SmartEP by Intelligent Hearing Systems, and Audix by Neuronic have successfully commercialized automated ASSR systems.

In addition to utilizing the amplitude and phase spectra of the FFT, ASSR automation can also be accomplished by evaluating polar vectors of the FFT results. For example, the AUDERA ERA system utilizes polar vectors to achieve ASSR automation. In this system, the underlying assumption is that patterns of the polar vectors obtained from a no response recording will likely have a random distribution. Generally speaking, the amplitude and phase of the polar vectors will be randomly distributed when noise is dominating a recording. On the other hand, when a response is present, the amplitude and phase of the polar vectors should have a uniform distribution. An example of this application can be seen in Figure 2.

An alternative method in ASSR automation, which bypasses the need for FFT, has been proposed. The RapidASSR works differently from other systems by implementing an adaptive filtering algorithm. This algorithm utilizes a Fourier Linear Combiner to adjust the amplitude and phase of the frequency bins that may exist in a conventional FFT analysis. It should be noted that this algorithm does not require an FFT analysis; instead, it utilizes an adaptive filtering algorithm. In this algorithm, once the test reaches a predetermined confidence interval, an auto stop feature is enacted and the ASSR is deemed present. If the level is not met in a recording after a designated acquisition time, then the recording is stopped and titled absent.

With the overarching goal of estimating hearing sensitivity at multiple frequencies, the success of ASSR automations has provided multiple tools that can be used in clinical and research applications. The automated ASSR is frequency specific and can be administered simultaneously at multiple frequencies in both ears. Such capabilities may be the reason some researchers are inclined to perform automated ASSR algorithms over ABR algorithms (Lins et al., 1996).

Figure 2

Schematic polar plots of ASSR recordings obtained in three hypothetical scenarios

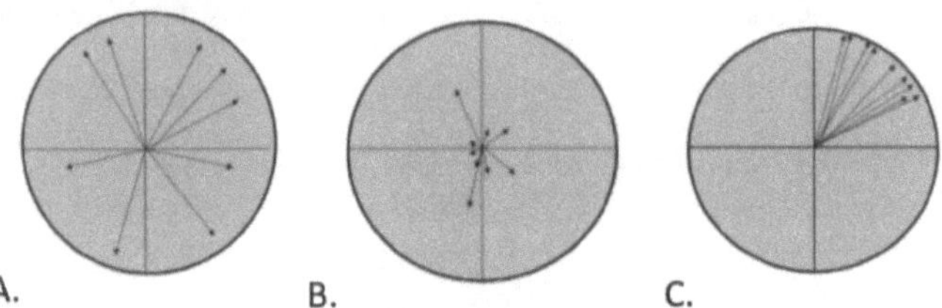

Panel A depicts a polar plot of a noisy subject who does not show a response (i.e., long vectors pointing at random directions), Panel E depicts a polar plot of a quiet subject who does not show a response (i.e., short vectors pointing at random directions), and Panel I depicts a polar plot of a subject who shows a response (i.e., vectors that are clustered together and pointed to a similar angle).

Preliminary Studies

Researchers have also sought to investigate tonal detection in FFR recordings. Llanos and colleagues (2017) showed successful outcomes using a Hidden Markov Model (HMM) to decipher between FFR recordings elicited by four Mandarin tones. Results from their study reported that, with the use of an HMM, FFR recordings could be distinguished with high accuracies. Additionally, FFR recordings obtained from individuals who have tonal language exposure were classified more accurately than those obtained from individuals without tonal language exposure. This study reports the potential in using machine learning for detecting FFRs, as well as its potential diagnostic capabilities. In line with the findings from Llanos and colleagues, three pilot studies were completed in our laboratory.

Study 1: Machine learning in Detecting FFRs in Chinese Adults:

Our piloting research began with utilizing previous FFR recordings that were obtained from 25 Chinese adults, in response to the Mandarin monosyllabic /i/ tone with a rising frequency contour. In this pilot study, in order to produce a sufficient number of recording waveforms, all the EEG sweeps within a recording were iterated to create a total of 1500 sub-averaged waveforms. Six response key features (Frequency Error, Slope Error, Tracking Accuracy, Spectral Amplitude, Pitch Strength and Root-Mean-Square Amplitude) were then extracted from each of the 37500 sub-averaged waveforms (1500 sub-averaged waveforms/recording x 25 recordings/participant = 37500 sub-averaged waveforms). From there, recordings were divided into FFR-absent or FFR-present categories. Table 1 displays the outcomes of this pilot study after the data were fed into 23 machine learning algorithms available in MATLAB's Classification Learner Application. Sensitivity, specificity, false negative rate, false positive rate, and efficiency results were computed based off the outputs from the 23 machine-learning algorithms (Table 1). Among these algorithms, the Logistic Regression and Bagged Tree produced 100% sensitivity and 100% specificity, and thus 100% efficiency. It is important to note that a total of 16 machine learning algorithms demonstrated ≥ 99% efficiency. In summary, this pilot study demonstrated that machine learning algorithms could be used to detect the presence of an FFR (Hart & Jeng, 2018).

Table 1.

Performance of 23 machine learning algorithms

Machine-Learning Algorithm	Sensitivity	Specificity	False Negative	False Positive	Efficiency
1. Complex Tree	99.84%	99.83%	0.16%	0.17%	99.83%
2. Medium Tree	99.84%	99.83%	0.16%	0.17%	99.83%
3. Simple Tree	99.78%	98.15%	0.22%	1.85%	99.15%
4. Linear Discriminant	88.10%	89.68%	11.90%	10.32%	88.71%
5. Quadratic Discriminant	96.77%	99.91%	3.23%	0.09%	97.99%
6. Logistic Regression	100.00%	100.00%	0.00%	0.00%	100.00%
7. Linear SVM	100.00%	99.91%	0.00%	0.09%	99.97%
8. Quadratic SVM	100.00%	99.74%	0.00%	0.26%	99.90%
9. Cubic SVM	100.00%	99.83%	0.00%	0.17%	99.93%
10. Fine Gaussian SVM	100.00%	99.83%	0.00%	0.17%	99.93%
11. Medium Gaussian SVM	100.00%	99.23%	0.00%	0.77%	99.70%
12. Coarse Gaussian SVM	100.00%	96.93%	0.00%	3.07%	98.81%
13. Fine KNN	100.00%	99.83%	0.00%	0.17%	99.93%
14. Medium KNN	100.00%	99.40%	0.00%	0.60%	99.77%
15. Coarse KNN	99.30%	93.09%	0.70%	6.91%	96.90%
16. Cosine KNN	100.00%	98.80%	0.00%	1.20%	99.54%
17. Cubic KNN	100.00%	99.49%	0.00%	0.51%	99.80%
18. Weighted KNN	100.00%	99.66%	0.00%	0.34%	99.87%
19. Boosted Trees	100.00%	0.00%	0.00%	100.00%	61.31%
20. Bagged Tree	100.00%	100.00%	0.00%	0.00%	100.00%
21. Subspace Discriminant	88.75%	83.11%	11.25%	16.89%	86.56%
22. Subspace KNN	100.00%	99.23%	0.00%	0.77%	99.70%
23. RUS Boosted Trees	99.84%	69.80%	0.16%	30.20%	88.21%

SVM: Support vector machine; KNN: K-nearest neighbors

Study 2: Machine learning in Detecting FFRs in American Neonates:

The purpose of this pilot study was to investigate whether results similar to Hart and Jeng (2018) could be repeated when utilizing recordings from neonates. In order to test the hypothesis of whether machine learning can detect FFRs from neonates, we proceeded to test American neonates (1-3 days after birth). FFR recordings were obtained with the same parameters as our previous pilot study. Recordings were obtained from 43 American neonates to evaluate the sensitivities, specificities, and efficiencies of multiple machine learning algorithms. FFR testing in was conducted in a newborn nursery center. Continuous brain waves were recorded in response to a pre-recorded monosyllable /i/ with a rising frequency contour from all neonates. A total of 8000 artifact-free sweeps were obtained from each participant.

Recordings were categorized into response and no response categories. Six response features (Frequency Error, Slope Error, Tracking Accuracy, Spectral Amplitude, Pitch Strength and Root-Mean-Square Amplitude) were extracted from each recording and served as key predictors in the identification of an FFR. Twenty-three supervised machine learning algorithms, with a 10-fold cross-validation procedure, were implemented via a Classification Learner Application in MATLAB to evaluate sensitivity, specificity, and efficiency. The mean efficiency values of the 23 algorithms were 86.0, 94.4, 97.2, and 97.5% when 1, 10, 100, and 1000 random iterations were implemented. Results of this study indicated that, through the combined efforts of supervised machine learning, iterations, and the six key features, the identification of the presence or absence of a recording could be obtained with high sensitivities, specificities, and efficiencies. These findings could serve as a foundational step in the forward progression of future studies in the detection of neonatal pitch processing (Hart & Jeng,

2020).

Table 2.

Performance outcomes (sensitivity, specificity, and efficiency) of the 23 machine learning algorithms when recordings went through 1, 10, 100, 1000 iterations.

Machine-Learning Algorithm	1 Iteration			10 Iterations			100 Iterations			1000 Iterations		
	Sensitivity	Specificity	Efficiency	Sensitivity	Specificity	Efficiency	Sensitivity	Specificity	Efficiency	Sensitivity	Specificity	Efficiency
1. Fine Tree	75.0%	92.8%	85.4%	91.0%	95.4%	93.5%	97.2%	97.9%	97.6%	98.3%	98.2%	98.2%
2. Medium Tree	75.0%	92.8%	85.4%	91.0%	95.4%	93.5%	97.7%	95.7%	96.6%	97.3%	95.5%	96.3%
3. Coarse Tree	75.0%	92.8%	85.4%	89.0%	92.9%	91.2%	95.9%	93.6%	94.5%	92.6%	95.3%	94.1%
4. Linear Discriminant	100%	89.3%	93.8%	98.5%	93.6%	95.6%	97.9%	94.9%	96.2%	97.5%	94.5%	95.8%
5. Quadratic Discriminant	80.0%	89.3%	85.4%	98.0%	95.7%	96.7%	97.5%	95.0%	96.0%	97.2%	94.7%	95.8%
6. Logistic Regression	85.0%	89.3%	87.5%	97.0%	98.2%	97.7%	97.0%	96.9%	96.9%	96.7%	96.6%	96.6%
7. Linear SVM	90.0%	89.3%	89.6%	98.5%	95.4%	96.7%	97.7%	96.5%	97.0%	97.2%	96.4%	96.7%
8. Quadratic SVM	90.0%	85.7%	87.5%	98.0%	97.5%	97.7%	98.9%	97.6%	98.1%	98.5%	98.1%	98.3%
9. Cubic SVM	85.0%	89.3%	87.5%	98.5%	98.2%	98.3%	99.0%	98.9%	99.0%	99.3%	99.4%	99.3%
10. Fine Gaussian SVM	60.0%	92.9%	79.2%	94.5%	97.9%	96.5%	98.9%	99.0%	99.0%	99.5%	99.6%	99.5%
11. Medium Gaussian SVM	90.0%	89.3%	89.6%	97.5%	96.4%	96.9%	99.3%	97.5%	98.3%	99.4%	99.1%	99.2%
12. Coarse Gaussian SVM	95.0%	89.3%	91.7%	92.5%	91.7%	92.1%	98.1%	95.2%	96.4%	98.7%	96.4%	97.3%
13. Fine KNN	90.0%	85.7%	87.5%	98.0%	96.8%	97.3%	98.7%	98.9%	98.8%	99.0%	99.1%	99.1%
14. Medium KNN	100%	89.3%	93.8%	95.5%	95.7%	95.6%	99.2%	97.7%	98.5%	99.1%	99.2%	99.2%
15. Coarse KNN	0%	100%	58.3%	96.0%	89.6%	92.3%	98.4%	95.0%	96.3%	98.8%	98.2%	98.4%
16. Cosine KNN	95.0%	89.3%	91.7%	97.0%	96.1%	96.5%	99.2%	98.0%	98.4%	99.0%	99.2%	99.1%
17. Cubic KNN	100.0%	89.3%	93.8%	95.5%	95.7%	95.6%	99.1%	98.0%	98.4%	99.1%	99.2%	99.2%
18. Weighted KNN	95.0%	89.3%	91.7%	98.5%	96.1%	97.1%	99.2%	98.2%	98.6%	99.3%	99.2%	99.2%
19. Boosted Trees	0%	100%	58.3%	37.0%	98.9%	73.1%	98.8%	97.8%	98.2%	99.1%	97.7%	98.3%
20. Bagged Trees	85.0%	89.3%	87.5%	95.5%	96.1%	95.8%	98.0%	98.3%	98.2%	98.9%	99.2%	99.0%
21. Subspace Discriminant	95.0%	89.3%	91.7%	96.5%	91.1%	93.3%	95.4%	92.0%	93.4%	95.3%	91.8%	93.3%
22. Subspace KNN	100%	85.7%	91.7%	91.0%	93.9%	92.7%	93.2%	94.8%	94.2%	94.2%	95.4%	94.9%
23. RUS Boosted Trees	65.0%	96.4%	83.3%	93.5%	95.4%	94.6%	98.1%	95.6%	96.6%	97.6%	95.7%	96.5%

SVM: Support vector machine; KNN: K-nearest neighbors

Study 3: Feasibility of Noise Separation in FFRs by using the NMF Algorithm

In this feasibility study, NMF procedures were applied to an exemplary FFR recording. One neonatal recording from the previous pilot study was utilized for subsequent NMF analysis. For this exemplary recording, continuous brainwaves were recorded from the monosyllable /i/ with a rising frequency contour for a total of 8000 artifact free sweeps. In order to produce a sufficient number of sub-averaged waveforms, the following procedure was administered. First, 500 sweeps were randomly selected from all available sweeps to constitute a sub-averaged waveform. Second, this procedure was repeated 5 times (nIterations). All sub-averaged waveforms were converted into amplitude spectrograms, which were fed into the NMF algorithm to compute the W (noise) and H (targeted response) components of each sub-averaged waveform. Within the NMF algorithm, the W and H matrices were repeatedly adjusted for a total of 100 times (nRepetitions) for the best optimization results. In ideal situations, the H component would contain nothing but the targeted response (i.e., its amplitude spectrum would display spectral energies that closely follow the frequency contour of the stimulus) and the W component would contain nothing but noise (i.e., its amplitude spectrum would display spectral energies that are randomly distributed). This process was then repeated to randomly select 1000, 2000, 3000, 4000, 5000, 6000, 7000, and 8000 sweeps. In this specific scenario, the sub-average of 8000 sweeps was treated as a reference recording. The reference recording served as a baseline for comparative measures.

The output of the NMF algorithm were two amplitude spectrograms derived from the W and H components. While the amplitude spectrogram of W is representative of noise, the amplitude spectrogram of H is indicative of the targeted response. In this feasibility study, Root Mean Square Errors (RMSE) were derived from each amplitude spectrogram of H that was obtained from a sub-averaged waveform and the amplitude spectrogram of H that was obtained from the referenced waveform. The same procedure was applied to all sub-averaged waveforms. Because the RMSE represents the difference between the amplitude spectrogram of H derived from a sub-averaged waveform and that of a referenced waveform, the RMSE value was considered a valid measurement of the performance of the NMF algorithm.

It should be noted that, addition to the NMF algorithm, the SNR of a recording could be increased simply by increasing the number of sweeps included in a sub-averaged waveform. Such a confounding factor should be removed. In order to remove the effects of signal averaging from those of the NMF algorithm, an additional "control" condition was administered. Specifically, each sub-averaged waveform was generated as previously mentioned, but the NMF algorithm was not administered. An RMSE value was then computed as the difference between the amplitude spectrum of a sub-averaged waveform and that of a referenced waveform. Because the NMF algorithm was not administered, this additional control condition provided a method to quantify the effects of signal averaging alone. Importantly, this control condition allows the experimenter to derive the effects the NMF algorithm, without the contamination of the effects of signal averaging. In other words, the RMSE value derived from the experimental condition where both the NMF algorithm and signal averaging were administered would contain effects of both the NMF algorithm and signal averaging. On the other hand, the RMSE value derived from the control condition where only signal averaging was administered would contain only the effects of signal averaging. As a result, the difference between these two RMSE values would reveal the true effects of the NMF algorithm.

The same procedures were applied to the Correlation Coefficient (CC). A CC value was computed and defined as the correlation coefficient between the amplitude spectrogram of H derived from a sub-averaged waveform and that from a referenced waveform. The experimental condition was conducted when both the NMF algorithm and signal averaging were administered, and a control condition was performed when only the signal averaging was performed. The difference of the two CC values derived from the experimental and control conditions would be considered the true effects of the NMF algorithm.

Results of this feasibility study demonstrated that noise separation could be accomplished when as few as 500 sweeps were included in a sub-averaged waveform. It should be noted that the RMSE values produced when only 500 sweeps were included (and the NMF algorithm was applied) were similar to those produced from a recording when 3000 sweeps were included (but the NMF algorithm was not applied). When the NMF algorithm is applied, a smaller number of sweeps was needed to produce a similar RMSE value, when compared to a conventional signal averaging without NMF. The CC values displayed similar, albeit opposite, results to the RMSE values. The CC values produced similar results. When as few as 500 sweeps were included when the NMF algorithm was applied, its CC values were similar to those produced 3000 sweeps when the NMF algorithm was not applied. In summary, with the NMF algorithm applied, the inclusion from as few as 500 sweeps produced comparable results to a conventional averaging of 3000 sweeps (Jeng et al, unpublished data).

Figure 3

Outcomes from the NMF Feasibility study

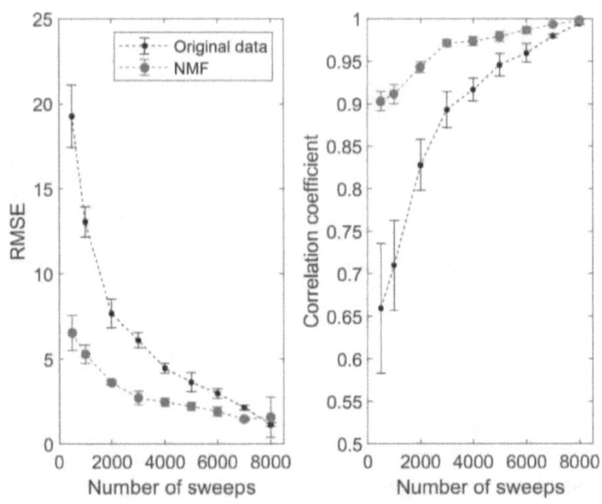

Root Mean Square Error (RMSE) and Correlation Coefficient (CC) values that were

obtained when the NMF algorithm was applied (i.e., the experimental condition, marked

in red) versus when the NMF algorithm was not applied (i.e., the control condition,

marked in black). Each vertical bar indicates one standard deviation.

Chapter 3: Methodology

Participants

Due to the current restrictions in recruitment processes and face to face contact

with potential patients, preexisting data were utilized from our lab for this study.

Continuous brain waves were obtained from 30 American adults (27 females and 3

males, 18-40 years old) and 30 American Neonates (13 girls and 17 boys, 1-3 day old).

All participants had normal hearing sensitivity, which is less than or equal to 20 dB HL

from 125-8000 Hz. Hearing sensitivity for the neonates was measured through

otoacoustic emissions or ABR screenings.

Recording and Stimulation Parameters

The stimulus utilized for testing was the English vowel /i/, with a rising frequency

contour. The stimulus had a duration of 150 ms (or 250 ms, depending on the existing

databases in our lab) and a 10-ms cosine-square ramping at both ends. The stimulus was

presented monaurally at 70 dB SPL, via an electromagnetically shielded ER-3A insert

earphone, with a silent interval of 45 ms between adjacent stimuli. A minimum of 8000

accepted sweeps (maxSweeps) were obtained from both the adult and neonatal testing

conditions to aid in subsequent analysis.

For the adult recording parameters, three gold-plated electrodes were placed

below the hairline of the high forehead (non- inverting), right mastoid (inverting), and

low forehead (ground) on the participants' heads. In the neonatal recordings, three snap-

on electrodes (Ambu Neuroline 720) were placed below the hairline of the high forehead

(non-inverting), right mastoid, and left mastoid on the neonate's head. The inverting and

ground electrodes were determined based on the results of the neonate's hearing screening. If the neonate passed the hearing screen in both ears, the inverting electrode was placed on the right mastoid. If the neonate only passed in the left ear, the inverting electrode was placed on the left mastoid. All electrode impedances for the recordings obtained from the adult and neonatal participants were kept under 5000 Ohms. Brainwave recordings were amplified (Opti Amp 8002, gain 100000), bandpass filtered (10-3000 Hz, slope 6 dB/octave), and digitized at a rate of 20000 samples/s. Stimulus presentation and recording parameters were controlled by custom-made software written in LabView (National Instruments, Austin, USA).

Data Analysis

Offline data analysis was completed through custom MATLAB (MathWorks, Natick, MA) scripts. Initial preprocessing of the data occurred in a fixed order. All signals were first band pass filtered from 100-1500 Hz with a 500th order, to guarantee that data analysis focused predominately on the frequency of interest. After filtering, the data was converted into sweeps. A baseline correction was then applied to ensure that data estimation was conducted at a standardized zero-baseline level. From there, artifact rejection criteria were implemented, and an individual sweep was accepted if the restricted voltage did not surpass ±25 µV. Finally, the accepted sweeps that had passed the initial signal processing procedures were stored on a computer and were used for subsequent data analysis.

Extraction of FFR Features

FFR features such as Frequency Error, Slope Error, and Tracking Accuracy were computed to further evaluate the effects of the NMF algorithm on these measurements. The explanations and calculations of these FFR features are defined as follows.

1. Frequency Error is a measure of the accuracy of pitch-encoding throughout the stimulus presentation. This measure was calculated by obtaining the difference between the stimulus $f0s$ and response $f0s$. In this measure, the computation began with defining a frequency range in which the $f0s$ of a recording were picked out across the frequency spectrum. This method of picking the highest $f0$ data point per time bin was conducted across the 201 time bins. From there, once all points were determined in both the stimulus and response domain, the difference between the two were calculated to determine the frequency error for a recording.

2. Slope Error is a representation of the brainstem's ability to preserve the overall shape of the pitch contour of the stimulus signal. This measure was calculated by obtaining the difference of the slope estimates between the stimulus and response $f0$ contours. The process of determining the $f0s$ for the stimulus and response were conducted in the same manner as the frequency error. From there, slope estimates were generated based upon those $f0$ curves of the stimulus and the response. Once established, the slope error was calculated by subtracting the slope of the stimulus from the slope of the response.

3. Tracking Accuracy reflects how accurate pitch tracking is at the subcortical level. This feature was computed as the regression r value between the stimulus and the response $f0$ contours. This function was displayed as a measurement of the stimulus $f0$ contour on the x-axis and the response $f0$ contour plotted on the y-axis. In the simplest of terms, the

straighter the regression line generated between the two axes, the better the correlation of how well pitch tracking is encoded at the level of the subcortex.

Non-negative Matrix Factorization

The utilization of the NMF algorithm was applied to the raw data (i.e., amplitude spectrograms) obtained from the participant recordings. Custom NMF scripts written in MATLAB were implemented to analyze noise separation in FFRs, as well as uncover the "hidden" portion of the recorded response. To reiterate the equation for the NMF algorithm, V represents the original participant recording, W represents the noise in the recording, and H represents the targeted response in the recording. Ideally, after NMF application, the H component should display spectral energies that closely follow the frequency contour of the stimulus. The two amplitude spectrograms derived from W and H, as well as the amplitude spectrogram derived from V, aided in parameter manipulation and subsequent data analysis.

As previously mentioned, parameters such as maxSweeps, nSweeps, and nIterations, were manipulated in order to test the hypotheses for Aims 1 and 3. In sum, the maxSweeps parameter was defined as the maximum number of sweeps obtained in a recording session for subsequent analysis (i.e., 8000 sweeps). The nSweeps parameter was described as the number of sweeps used out of the total sweep count for data analysis (e.g., to include only 500 sweeps out of the total number of sweeps that are available for analysis). The nIterations parameter served to simulate higher volumes of participant data. Due to the current restrictions, this study only utilized data from 60 subjects. When 10 iterations were applied, data from 60 participants simulated results similar to recordings obtained by 600 individuals. Through the use of the NMF algorithm and the manipulations of the three parameters, FFR researchers could be able to develop a method which could identify FFR recordings in a timely manner.

The three FFR features were extracted from the amplitude spectrograms before and after the NMF algorithm was applied to each recording. The FFR features derived in the experimental and control conditions were plotted to illustrate the effects of the NMF algorithm at various maxSweeps, nSweeps, and nIterations manipulations. Graphical presentations were also generated to demonstrate the trend and differences between the two participant groups.

Statistical Analysis

In order to examine the NMF algorithm's effects on FFR recordings, three separate statistical analysis were conducted for each specific aim. Although each of the parameters (maxSweeps, nSweeps, and nIterations) could range from one to infinity, it was impossible to accomplish that in reality. As a result, a list of selected numbers was utilized in order to encompass a full range of maxSweeps, nSweeps, and nIterations for thoroughness. For this study, the NMF parameters were manipulated as follow:

- maxSweeps: 8000

- nSweeps: 100, 250, 500, 1000, 2000, 3000, 4000, 5000, 6000, 7000, 8000

- nIterations: 1, 5, 10, 100

Aim 1

First, a two-way ANOVA was utilized to investigate the significance within and across the two main factors (main factor 1. The NMF algorithm [i.e., NMF + Signal Averaging {NMF condition} or just Signal Averaging {ORI condition}] and 2. nSweeps condition [i.e., 100, 250, 500. 1000, 2000, 3000, 4000, 5000, 6000, 7000, 8000 sweeps])

for the adult FFRs. Post-hoc procedures with a Bonferroni adjustment were conducted to

assess the significance of all pair-wise comparisons.

Second, a linear regression was utilized for statistical analysis to understand the

overall incremental and decremental trends of the FFR. An assessment of the NMF

parameters (maxSweeps, nSweeps, and nIterations) were made in relation to each of the

FFR features (Frequency Error, Slope Error, and Tracking Accuracy). Such assessment

was performed for both the ORI and NMF conditions.

Third and lastly, exponential modeling was used to better delineate the

incremental or decremental trends of pitch processing in the brainstem as a function of

sweeps for aims 1. This measure aided in the understanding and quantification for timing

purposes of the recording.

To calculate frequency error, which has a descending trend with increasing

number of sweeps, the following model was used:

$$A(n) = A_{noise}\left(e^{-\frac{n}{t}}\right) - A_{AS} \tag{5}$$

For slope error and tracking accuracy, which has an ascending trend with

increasing number of sweeps, an alternative model was used:

$$A(n) = A_{AS}\left(1 - e^{-\frac{n}{t}}\right) - A_{noise} \tag{6}$$

For clarity, A is an objective measure (i.e., *Frequency Error or Slope Error*) of

the FFR to voice pitch. n is the number of sweeps included in the averaging process.

A_{noise} represents the amplitude of noise and was derived from the fitted curve of the FFR

trend of a specific objective index when the number of sweeps equals 100 (i.e., units of

A_{noise} remain the same for each of the five objective indices). A_{AS} represents the asymptotic amplitude of the response and was computed from the fitted curve of the exponential model with the number of sweeps being 8000. e is Euler's number: 2.7182; τ is the 'time' constant of the fitted curve that denotes the number of sweeps needed to reach its 63% asymptotic amplitude. Calculation and derivation of the 63% asymptotic amplitude was based on the mathematical principle that the exponential function, with a base of e, was identical to its derivative (Courant & Robbins, 1996; Goldstein et al, 2009).

Aim 2

A one-way ANOVA was utilized to investigate the research question for aim 2. In order to understand the effects of the hidden component, uncovered by the NMF algorithm, this analysis measure was implemented. Post-hoc procedures with a Bonferroni adjustment were conducted to assess the significance of all pair-wise comparisons. Similarly, a linear regression analysis was utilized to better understand the incremental and decremental trends of the hidden component for an adult FFR. An assessment of the NMF parameters (maxSweeps, nSweeps, and nIterations) were made in relation to each of the Hidden FFR features (hidden Frequency Error [hFE], hidden Slope Error[hSE], and hidden Tracking Accuracy [hTA]). Finally, exponential modeling was implemented to address the differing trends of the "hidden" portion of the response. As well as to investigate and quantify the timing of the hidden response.

Aim 3

Beginning aim 3, a two-way ANOVA was utilized to investigate the significance within and across the two main factors (main factor 1. The NMF algorithm [i.e., NMF +

Signal Averaging {NMF condition} or just Signal Averaging {ORI condition}] and 2. nSweeps [i.e., 100, 250, 500. 1000, 2000, 3000, 4000, 5000, 6000, 7000, 8000 sweeps]) for the newborn FFRs. Post-hoc procedures with a Bonferroni adjustment were conducted to assess the significance of all pair-wise comparisons. Next, a linear regression analysis was utilized to understand the incremental and decremental trends for the newborn FFR. Finally, to better investigate and quantify the trends and timing of the newborn FFR, exponential modeling was implemented.

Then the same procedure was applied to further investigate the significance of the three main factors (1. Age groups [i.e., adults versus neonates], nSweeps condition [100, 250, 500, 1000, 2000, 3000, 4000, 5000, 6000, 7000, and 8000 sweeps], and the NMF algorithm [NMF condition or ORI condition]) for the age comparisons of the recording. Then, a linear regression analysis was utilized to understand the incremental and decremental trends for the age differences. Lastly, to better delineate the incremental or decremental trends of pitch processing in the brainstem as a function of sweeps exponential modeling was conducted.

Aim 4

In order to understand the effects of the newborn hidden component, uncovered by the NMF algorithm, a one-way ANOVA measure was implemented. Post-hoc procedures with a Bonferroni adjustment were conducted to assess the significance of all pair-wise comparisons. In line with aim 2, a linear regression analysis was applied to better understand the incremental and decremental trends of the hidden component for a newborn FFR. An assessment of the NMF parameters (maxSweeps, nSweeps, and nIterations) were made in relation to each of the hidden FFR features (hFE, hSE, and hTA). Finally, exponential modeling was implemented to address the differing trends of the "hidden" portion of the response. As well as uncover and understand the timing trends of the newborn hidden response.

Then the same procedure was applied to further investigate the significance of the two main factors (1. Age groups [i.e., adults versus newborns] and the nSweeps condition [100, 250, 500, 1000, 2000, 3000, 4000, 5000, 6000, 7000, and 8000 sweeps]) for the age comparisons of the hidden recording. Then, a linear regression analysis was utilized to understand the incremental and decremental trends for the hidden age differences. Lastly, to better delineate the incremental or decremental trends of the hidden response exponential modeling was conducted.

Chapter 4: Results

Noise Separation by Using the NMF Algorithm

Results from the NMF separation can be seen in Figures 4 and 5. Beginning with Figure 4, this figure displays a typical adult recording (Adult005). Results of the NMF separation can be viewed in 4 separate rows. The first row shows the original FFR spectrograms across the 11 nSweeps conditions, best described as the V portion of the NMF equation (V = W x H). This row plots the original FFR spectrograms that contain both the response and noise in the recordings. Row 2 shows the number of sweeps used in the averaging procedure, ranging from 100-8000 sweeps. Row 3 shows the separated FFR (H portion of the NMF equation). This row shows ideally the targeted neural response of the recording, uncovered by the NMF algorithm. This row displays results for each of the nSweeps tested for separation (1=100, 2=250, 3=500, 4=1000, 5 = 2000, 6 = 3000, 7 = 4000, 8 = 5000, 9 = 6000, 10 = 7000, & 11 = 8000 sweeps). Finally, the last row depicts the W portion of the NMF equation, which is noted as the separated noise component of each recording.

When viewing Figure 4 as a whole, multiple observations can be made. Beginning with row 1, results were consistent with a conventional averaging procedure (without the use of the NMF algorithm) in a typical adult recording. The spectrogram of the first 100 sweeps showed no visually discernible spectral energies that could be definitively determined as a response. As the number of sweeps increased, the spectral energies at the fundamental frequency and its harmonics became better defined (through a visual inspection) and showed a mirrored depiction of the rising frequency contour of the stimulus.

When comparing the results of row 1 (original spectrograms) to row 3 (separated FFR spectrograms), it was more visually apparent that the spectral energies were enhanced along the fundamental frequency and its harmonics in row 3. There was also a visual enhancement of the first harmonic in row 3. With the NMF algorithm, the spectrogram output as a whole became well defined and better in terms of discernibility. At lower sweep levels, such as 500 and 1000 sweeps, discernible spectral energies were more apparent in row 3 (compared to row 1), and thus making a response more visually detectible.

In order to understand why there was an enhancement in row 3, one must look at row 4. Row 4 showed the amount of noise which was removed from the original recording. The extracted noise (indicated by the extent of color variations) could be observed mostly at the lower sweep counts. From there, the amount of extracted noise started to decrease with increasing number of sweeps and became relatively steady starting from about 6000 sweeps. The separated noise appeared to contain no relationship with the separated FFRs. These findings indicated that a substantial amount of noise was removed through the use of the NMF algorithm.

Figure 4

Graphical presentations of the NMF separation procedure for an adult participant.

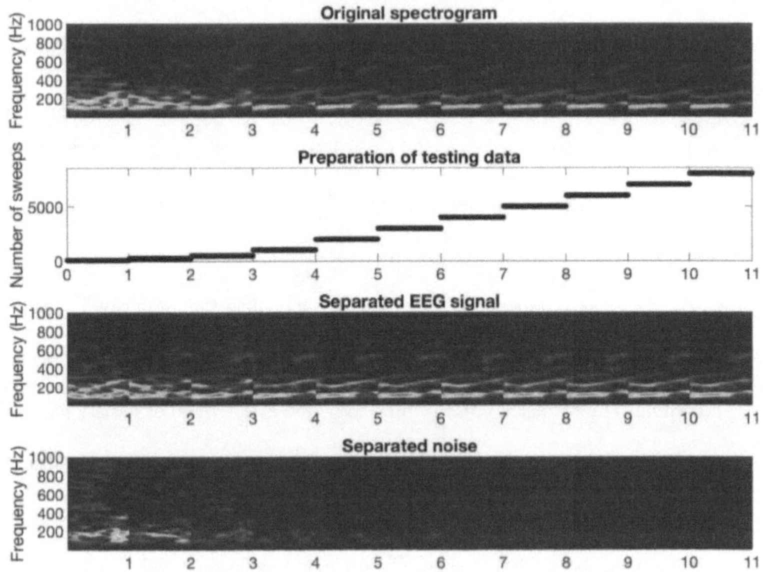

Row 1 depicts the amplitude spectrograms before the use of the NMF algorithm. Row 2 displays the selected number of sweeps utilized in the averaging procedure. Row 3 denotes the targeted neural response from the original spectrogram (H portion of the NMF equation) after the NMF algorithm is applied. Row 4 displays the noise component (W portion of the NMF equation) that has been removed from the original spectrogram. These results were obtained at 1 iteration.

Spectrograms obtained from a typical newborn recording (Newborn025) are displayed in Figure 5. When viewing Figure 5, there are similarities in the results as those shown in Figure 4. The newborn data showed a similar trend in FFR observability. For example, at lower sweeps, an FFR was not visually identifiable; additionally, as more sweeps were included in the recording, an FFR became visually identifiable. When comparing the number of sweeps required to obtain a visually identifiable FFR, it required a much higher number of sweeps (6000-8000 sweeps) in neonatal recordings than the number of sweeps (5000-6000 sweeps) needed for adult recordings. Through the use of the NMF algorithm, noise separation was also apparent in these neonatal recordings.

Figure 5

Graphical presentations of the NMF separation procedure for a newborn participant.

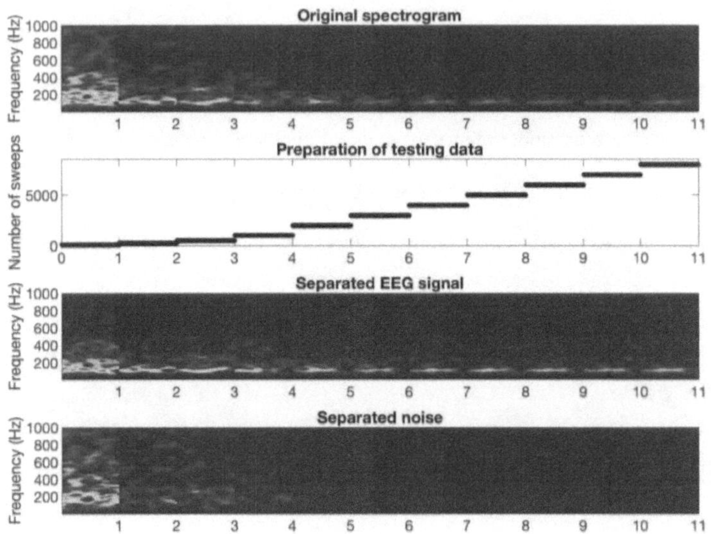

Row 1 depicts the amplitude spectrograms before the use of the NMF algorithm. Row 2 displays the selected number of sweeps utilized in the averaging procedure. Row 3 denotes the targeted neural response from the original spectrogram (H portion of the NMF equation) after the NMF algorithm is applied. Row 4 displays the noise component (W portion of the NMF equation) that has been removed from the original spectrogram. These results were obtained at 1 iteration.

The results of the NMF algorithm at 5, 10, and 100 iterations for the adult and newborn participants can be viewed in Appendices A and B, respectively. Similar comparisons and findings at 5, 10 and 100 iterations were also observed in reference to the 1 iteration findings. As nSweeps increased, response visibility increased when only the traditional averaging technique was used. With the NMF algorithm, there was an enhancement in FFR observability. Additionally, the number of sweeps required to obtain a visually detectable FFR was decreased substantially for both the adult and newborn participants when the NMF algorithm was applied.

Application of the NMF Algorithm in Adults (Aim 1)

To reiterate, the purpose of Aim 1 was to systematically manipulate NMF parameters and investigate their performances of noise separation from FFRs in adult participants.

Figure 6 depicts the results of the ORI and NMF conditions for adults. Panel A displayed the FE results. For 1 iteration, when the conventional signal averaging was utilized (i.e., before the NMF algorithm was applied), the FFR was 13.218 ± 0.801 Hz at 100 sweeps. With an increasing number of sweeps, FE values decreased and became relatively stable at roughly 6000-8000 sweeps. When the NMF algorithm was utilized, the FFR values followed a similar trend as those in the ORI condition, with a few differences. One difference was that the FE values at 100 Hz were smaller in the NMF condition compared to those of the ORI condition (7.583 ± 0.627 Hz for the NMF condition and 13.218 ± 0.801 Hz for the ORI condition). Another difference was that FE values reached a plateau at roughly 2000-4000 sweeps in the NMF condition. When comparing results of the two conditions (ORI & NMF), it was apparent that the NMF condition required a smaller number of sweeps in order to obtain comparable results to the ORI condition. Through visual inspections, there was an apparent improvement in the FE values when the NMF algorithm was implemented. When viewing the data from 5, 10, and 100 iterations (Panels B, C, & D), results were consistent with those observed at 1 iteration. The FE feature showed an improvement with the NMF condition when compared to the ORI condition.

The SE results are displayed in Panel E. When viewing the ORI data for 1 iteration, SE values were smallest at 100 sweeps (-213.467 ± 20.303 Hz/ms). With an increasing number of sweeps, SE values increased and became closer to zero. Approximately 5000-8000 sweeps were needed to reach a plateau in SE values. When visualizing the NMF data, as nSweeps increased, SE values increased and became closer to zero. One difference was that the SE values were smaller at 100 Hz in the NMF condition than those in the ORI condition (-96.156 ± 15.334 Hz/ms for the NMF condition and -213.467 ± 20.303 Hz/ms) for the ORI condition). Another difference was that roughly 2000-4000 sweeps were required to see a stabilization in SE values. There was an apparent improvement in the SE values when the NMF algorithm was implemented. Furthermore, the data from 5, 10, and 100 iterations (Panels F, G, & H) were in line with the 1 iteration results. SE values without the NMF algorithm showed a similar trend in improvement; however, when the NMF algorithm was implemented, SE values began at a higher SE value at 100 sweeps and improved quicker (i.e., a faster rate of improvement).

The TA feature can be viewed in Panel I. For the ORI condition at 1 iteration, TA values were smallest (0.371 ± 0.042) at 100 sweeps. As nSweeps increased, TA values increased and became relatively stable. Roughly 4000-8000 sweeps were needed to observe a plateau in FE values through conventional signal averaging. When looking at the NMF data, similarities were present between the ORI and NMF conditions. However, there were also differences in the TA values between the ORI and NMF conditions (0.711 ± 0.052 for the NMF condition and 0.371 ± 0.042 for the ORI condition). Another difference was that roughly 1000-4000 sweeps were required to see a plateau in TA features when the NMF algorithm was applied. It was apparent that the NMF condition required a smaller number of sweeps in order to obtain comparable results to the ORI condition. When viewing the data from 5, 10, and 100 iterations (Panels J, K, & L), results remained consistent with those observed at 1 iteration.

Figure 6

Graphical presentations of the raw data extracts for the ORI and NMF conditions for the adult participants at 1, 5, 10 and 100 iterations for Frequency Error (first row), Slope Error (second row), and Tracking Accuracy (third row).

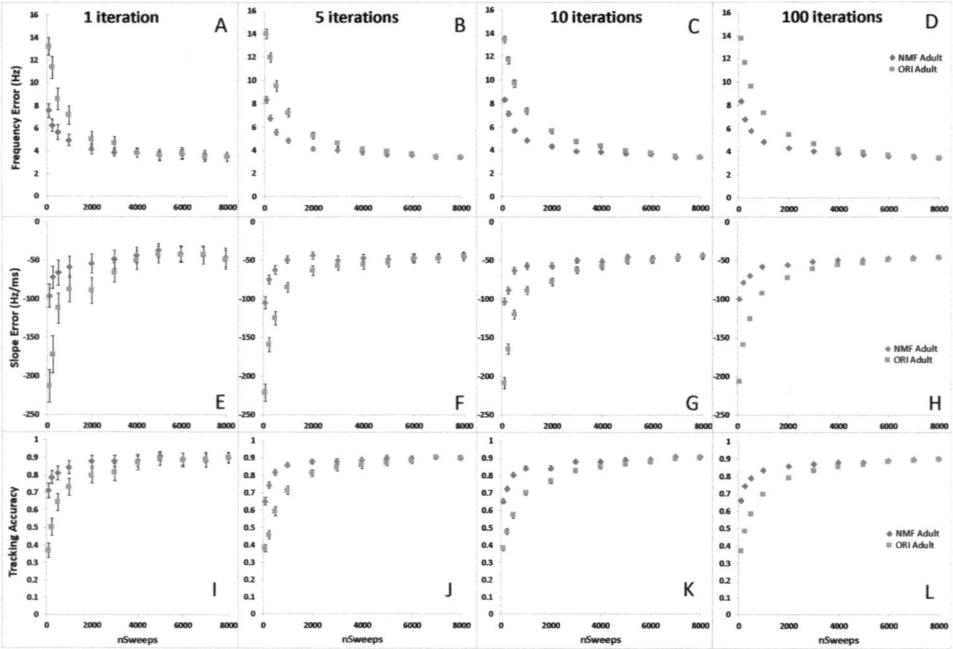

Linear Regression

Results of the linear regression analysis between the ORI and NMF adult data can be seen in Figure 7. Panel A depicts the results of the linear regression with the nSweeps condition on the x-axis and FE on the y-axis, Panel E depicts the results of the linear regression with the nSweeps condition on the x-axis and SE on the y-axis, and Panel I depicts the results of the linear regression with the nSweeps condition on the x-axis and TA on the y-axis. The four columns display results obtained at 1, 5, 10, and 100 iterations.

When viewing linear regression results, the slope of the regression indicates the rate of change or steepness of the line in the graph. The intercept is the indicator of where on the y-axis the line will cross when x is zero. Ideally speaking, the intercept of the equation represents the noise floor (i.e., the starting point) of the FFR trend. Generally speaking, all resultant linear regressions confirmed a decremental trend in the FE results and an incremental trend in the SE and TA results. For clarity, all resultant linear equations are grouped below by iterations.

1 Iteration.

$$\text{ORI Linear Regression for FE: } y = -0.001x + 9.5572 \tag{7}$$

$$\text{NMF Linear Regression for FE: } y = -0.0004x + 5.9102 \tag{8}$$

$$\text{ORI Linear Regression for SE: } y = 0.016x - 141.76 \tag{9}$$

$$\text{NMF Linear Regression for SE: } y = 0.0047x - 70.9 \tag{10}$$

$$\text{ORI Linear Regression for TA: } y = 5\text{E-}05x + 0.5826 \tag{11}$$

$$\text{NMF Linear Regression for TA: } y = 2\text{E-}05x + 0.7989 \tag{12}$$

For FE, slope and intercept estimates for Equations 7 and 8 displayed differing values. The ORI slope was steeper than that of the NMF condition. This meant that, for the ORI condition, FE values decreased at a faster rate compared to those in the NMF condition. In other words, the decrement for the ORI FE values took a smaller number of nSweeps (when compared with the NMF FE values) to obtain a similar performance. When viewing the differing intercepts, values for the NMF condition indicated a lower noise floor for the FE trend.

When looking at the SE results shown in Equations 9 and 10, there were different slope estimates. The ORI slope was steeper than that of the NMF condition. This meant that, for the ORI condition, SE values increased at a faster rate compared to those in the NMF condition. In other words, the decrement for the ORI SE values took a smaller number of nSweeps (than the NMF SE values) to obtain a similar performance. When viewing the different SE intercepts, values for the NMF condition indicated a lower noise floor for the SE trend.

For TA, the ORI slope (Equation 11) was steeper than the NMF slope (Equation 12). Simply put, the trend for the ORI condition took a smaller number of sweeps to reach a comparable performance to the NMF condition. This meant that, for the NMF condition, TA values increased at a faster rate when compared to the ORI condition. When viewing the differing intercepts, values for the ORI condition indicated a lower noise floor for the FE trend.

5 Iterations.

ORI Linear Regression for FE: $y = -0.0011x + 10.131$ $\hspace{2cm}$ (13)

NMF Linear Regression for FE: $y = -0.0004x + 6.1808$ $\hspace{2cm}$ (14)

ORI Linear Regression for SE: y = 0.0154x - 138.85 (15)

NMF Linear Regression for SE: y = 0.004x - 69.657 (16)

ORI Linear Regression for TA: y = 5E-05x + 0.5649 (17)

NMF Linear Regression for TA: y = 2E-05x + 0.7779 (18)

When viewing the 5 iteration equations, the values remained similar to the those

obtained at 1 iteration.

10 Iterations.

ORI Linear Regression for FE: y = -0.0011x + 10.079 (19)

NMF Linear Regression for FE: y = -0.0005x + 6.2932 (20)

ORI Linear Regression for SE: y = 0.0153x - 139.55 (21)

NMF Linear Regression for SE: y = 0.0052x - 76.421 (22)

ORI Linear Regression for TA: y = 5E-05x + 0.5553 (23)

NMF Linear Regression for TA: y = 2E-05x + 0.7608 (24)

When viewing the 10 iteration equations, the values remained similar to the those

observed at 1 and 5 iterations.

100 Iterations.

ORI Linear Regression for FE: y = -0.0011x + 10.053 (25)

NMF Linear Regression for FE: y = -0.0005x + 6.2556 (26)

ORI Linear Regression for SE: y = 0.015x - 138.63 (27)

NMF Linear Regression for SE: y = 0.0046x - 74.54 (28)

ORI Linear Regression for TA: y = 5E-05x + 0.5595 (29)

NMF Linear Regression for TA: y = 2E-05x + 0.766 (30)

When viewing the 100 iteration equations, the values remained similar to those observed at 1, 5, and 10 iterations.

Figure 7

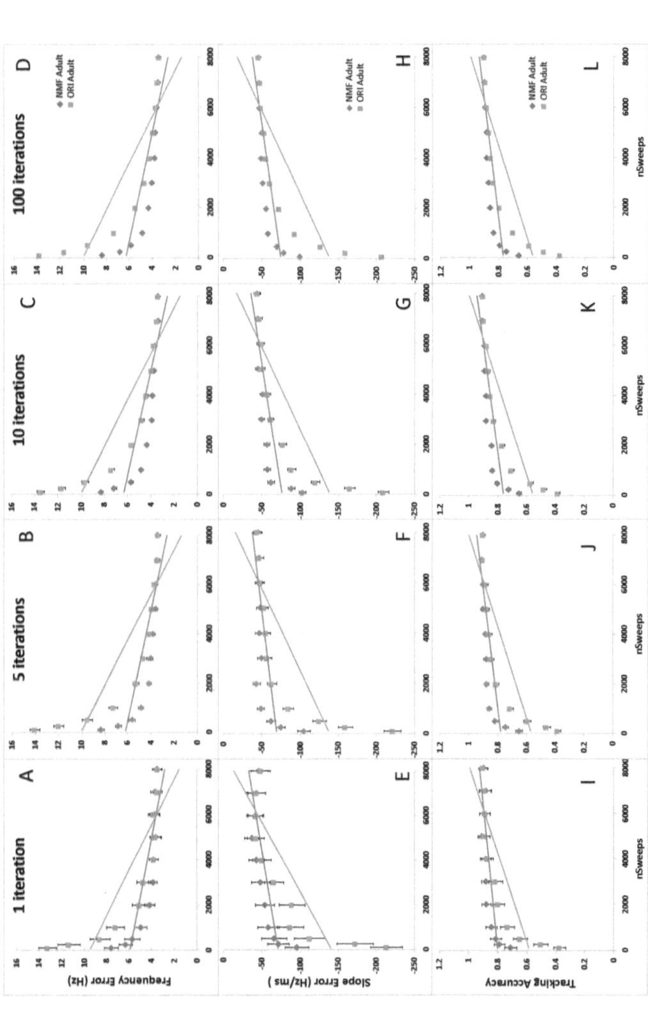

Graphical presentations of the linear regressions performed on the ORI and NMF conditions for the adult participants at 1, 5, 10 and 100 iterations for Frequency Error (first row), Slope Error (second row), and Tracking Accuracy (third row).

Exponential Modeling

Results of exponential modeling for the adult ORI and NMF conditions are displayed in Figure 8, respectively. Similar to the linear regression figures, Panel A depicts the FE results of exponential modeling, Panel E depicts the SE results of exponential modeling, and Panel I depicts the TA results of exponential modeling. Each column displayed results at 1, 5, 10, and 100 iterations, respectively.

The resultant exponential modeling equations are displayed below and grouped according to each nIterations condition. As previously described in the Methods section, A_{noise} represented the amplitude of noise and was derived from the fitted curve of the FFR trend of a specific objective index when the number of sweeps equaled 100. A_{AS} represented the asymptotic amplitude of the response and was computed from the fitted curve of the exponential model with the number of sweeps being 8000. The tau (τ) value provided an inclination estimate of the number of sweeps required for an FFR feature to reach its 63% asymptotic amplitude. Data from the exponential modeling confirmed a decremental trend in the FE results and an incremental trend in the SE and TA results. A tabular depiction of the Tau, A_{noise}, and A_{as} values can be viewed in Appendix G.

1 Iteration.

$$\text{ORI Exponential Modeling for FE: } A(n) = 12.635\left(e^{-\frac{n}{964.972}}\right) - 3.681 \tag{31}$$

$$\text{NMF Exponential Modeling for FE: } A(n) = 6.693\left(e^{-\frac{n}{1071.009}}\right) - 3.644 \tag{32}$$

$$\text{ORI Exponential Modeling for SE: } A(n) = -49.150\left(1 - e^{-\frac{n}{521.622}}\right) + 213.466 \tag{33}$$

$$\text{NMF Exponential Modeling for SE: } A(n) = -43.410\left(1 - e^{-\frac{n}{1107.723}}\right) + 77.357 \tag{34}$$

$$\text{ORI Exponential Modeling for TA: } A(n) = 0.887\left(1 - e^{-\frac{n}{824.472}}\right) - 0.425 \tag{35}$$

NMF Exponential Modeling for TA: $A(n) = 0.883\left(1 - e^{-\frac{n}{457.287}}\right) - 0.712$ $\qquad$ (36)

For FE, the exponential modeling Equations 31 and 32 displayed differing values. The noise amplitude (A_{noise}) values estimated for the ORI and NMF conditions were 12.635 and 6.693 Hz, respectively. These values indicated that the A_{noise} in the NMF condition was 47% ([12.635-6.693] / 12.635 = 47%) lower than that of the ORI condition. This demonstrated a 47% increment in the A_{noise} estimates of the NMF FE values (compared to the ORI FE values), and thus showing a better performance in the overall FE trends when the NMF algorithm was applied. The τ values of the fitted curves for the FE trends obtained in the ORI and NMF conditions were 964 and 1071 sweeps, respectively. It was important to note that the ORI condition produced a smaller τ value than the NMF. This finding indicated that the ORI condition had a faster improvement in frequency-encoding acuity (i.e., less frequency coding errors) with increasing number of sweeps compared to the NMF condition. This finding was surprising because the NMF algorithm was anticipated to produce a faster improvement. Such an unanticipated finding likely was due to the fact that the ORI FE values at 100 sweeps were potential outliers (i.e., larger than mean + 2 standard deviations) when compared to all the ORI FE values across the 11 nSweep conditions. The asymptotic amplitude (A_{AS}) values estimated for the ORI and NMF conditions were 3.681 and 3.644 Hz, respectively. Similar FE plateau values indicated that the NMF algorithm had little to no influence on the FE plateau values.

For SE, the exponential modeling Equations 31 and 32 displayed differing values. The asymptotic amplitude (A_{AS}) values estimated for the ORI and NMF conditions were -49.150 and -43.410 Hz/ms, respectively. Similar SE plateau values indicated that the

NMF algorithm had little to no influence on the SE plateau values. The τ values of the fitted curves for the SE trends obtained in the ORI and NMF participants were 521 and 1107 sweeps, respectively. It was important to note that the ORI condition produced a smaller τ value than the NMF condition. This finding indicated that the ORI condition showed a faster improvement in SE with increasing number of sweeps compared to the NMF. Similar to the FE values described in the previous paragraph, it should be noted that potential outliers were also observed for the SE values at 100 sweeps with the ORI condition, but not in the NMF condition. Such outliers could have adverse effects when estimating a τ value for the ORI condition. The noise amplitude (A_{noise}) values estimated for the ORI and NMF conditions were 231.446 and 77.357 Hz/ms, respectively. These values indicated that the A_{noise} values in the NMF condition were 66% ([{-231.446} – {-77.357}] / 231.446 = 66%) lower than those of the ORI condition. This 66% increment in the A_{noise} estimates of the SE values showed a better performance when the NMF algorithm was applied.

For TA, the exponential modeling Equations 33 and 34 displayed differing values. The asymptotic amplitude (A_{AS}) values estimated for the ORI and NMF conditions were 0.887 and 0.883, respectively. Similar TA plateau values indicated that the NMF algorithm had little to no influence on the TA plateau values. The τ values of the fitted curves obtained in the ORI and NMF participants were 824 and 457 sweeps, respectively. It was important to note that the adult participants produced a smaller τ value than neonates. This finding indicated that the NMF condition showed a faster improvement in pitch encoding (i.e., better tracking accuracy) with increasing number of sweeps compared to the ORI condition. The noise amplitude (A_{noise}) values estimated for the ORI and NMF conditions were 0.425 and 0.712, respectively. These values indicated that the A_{noise} values in the ORI condition were 40% ([0.712 - 0.425] / 0.712 = 40%) lower than those of the NMF condition. This 40% increment in the A_{noise} estimates of the TA values showed a better performance when the NMF algorithm was applied.

5 Iterations.

ORI Exponential Modeling for FE: $A(n) = 13.500\left(e^{-\frac{n}{863.430}}\right) - 3.847$ (37)

NMF Exponential Modeling for FE: $A(n) = 8.356\left(e^{-\frac{n}{463.151}}\right) - 3.585$ (38)

ORI Exponential Modeling for SE: $A(n) = -53.115\left(1 - e^{-\frac{n}{472.602}}\right) + 221.408$ (39)

NMF Exponential Modeling for SE: $A(n) = -47.048\left(1 - e^{-\frac{n}{256.498}}\right) + 104.963$ (40)

ORI Exponential Modeling for TA: $A(n) = 0.876\left(1 - e^{-\frac{n}{808.651}}\right) - 0.381$ (41)

NMF Exponential Modeling for TA: $A(n) = 0.896\left(1 - e^{-\frac{n}{350.057}}\right) - 0.651$ (42)

FE results, when viewing the 5 iterations, exponential modeling (Equations 37 and 38) displayed differing values. The τ values of the fitted curves for the FE trends obtained in the ORI and NMF conditions were 863 and 463 sweeps, respectively. It was important to note that the NMF condition produced a smaller τ value than the ORI. This finding indicated that the NMF condition showed a faster improvement in frequency-encoding acuity (i.e., less frequency coding errors) with increasing number of sweeps compared to the ORI condition. The FE A_{Noise} and A_{AS} values at 5 iterations were consistent with the 1 iteration results.

For SE, the exponential modeling Equations 39 and 40 displayed differing values. The SE τ values of the fitted curves for the ORI and NMF participants were 472 and 256 sweeps, respectively. It was important to note that the NMF condition produced a smaller τ value than the ORI condition. This finding indicated that the NMF condition showed a faster improvement in SE with increasing number of sweeps compared to the ORI condition. The SE A_{Noise} and A_{AS} values at 5 iterations were consistent with the 1 iteration results.

For TA, the exponential modeling Equations 41 and 42 displayed differing values. The τ values of the fitted curves for tracking accuracy obtained in the ORI and NMF conditions were 808 and 350 sweeps, respectively. It was important to note that the NMF condition produced a smaller τ value than the ORI condition. This finding indicated that the NMF condition showed a faster improvement in pitch encoding (i.e., better tracking accuracy) with increasing number of sweeps compared to the ORI condition. The TA A_{Noise} and A_{AS} values at 5 iterations were consistent with the 1 iteration results.

10 Iterations.

$$\text{ORI Exponential Modeling for FE: } A(n) = 13.008\left(e^{-\frac{n}{936.079}}\right) - 3.891 \tag{43}$$

$$\text{NMF Exponential Modeling for FE: } A(n) = 8.067\left(e^{-\frac{n}{648.570}}\right) - 3.670 \tag{44}$$

$$\text{ORI Exponential Modeling for SE: } A(n) = -50.722\left(1 - e^{-\frac{n}{487.249}}\right) + 208.657 \tag{45}$$

$$\text{NMF Exponential Modeling for SE: } A(n) = -47.028\left(1 - e^{-\frac{n}{499.799}}\right) + 103.203 \tag{46}$$

$$\text{ORI Exponential Modeling for TA: } A(n) = 0.870\left(1 - e^{-\frac{n}{896.119}}\right) - 0.406 \tag{47}$$

$$\text{NMF Exponential Modeling for TA: } A(n) = 0.890\left(1 - e^{-\frac{n}{417.735}}\right) - 0.652 \tag{48}$$

When viewing the 10 iteration equations, the values remained similar to those observed at 1 and 5 iterations.

100 Iterations.

ORI Exponential Modeling for FE: $A(n) = 12.983\left(e^{-\frac{n}{925.653}}\right) - 3.895$ $\qquad$ (49)

NMF Exponential Modeling for FE: $A(n) = 7.743\left(e^{-\frac{n}{700.376}}\right) - 3.721$ $\qquad$ (50)

ORI Exponential Modeling for SE: $A(n) = -53.037\left(1 - e^{-\frac{n}{760.014}}\right) + 182.605$ $\quad$ (51)

NMF Exponential Modeling for SE: $A(n) = -48.847\left(1 - e^{-\frac{n}{440.170}}\right) + 99.873$ $\quad$ (52)

ORI Exponential Modeling for TA: $A(n) = 0.868\left(1 - e^{-\frac{n}{918.529}}\right) - 0.414$ $\qquad$ (53)

NMF Exponential Modeling for TA: $A(n) = 0.881\left(1 - e^{-\frac{n}{444.510}}\right) - 0.661$ $\qquad$ (54)

When viewing the 100 iteration equations, the values remained consistent with those observed at 1, 5, and 10 iterations.

Figure 8

Graphical displays of the exponential modeling performed on the adult ORI and NMF data at 1, 5, 10 and 100 iterations for Frequency Error (first row), Slope Error (second row), and Tracking Accuracy (third row).

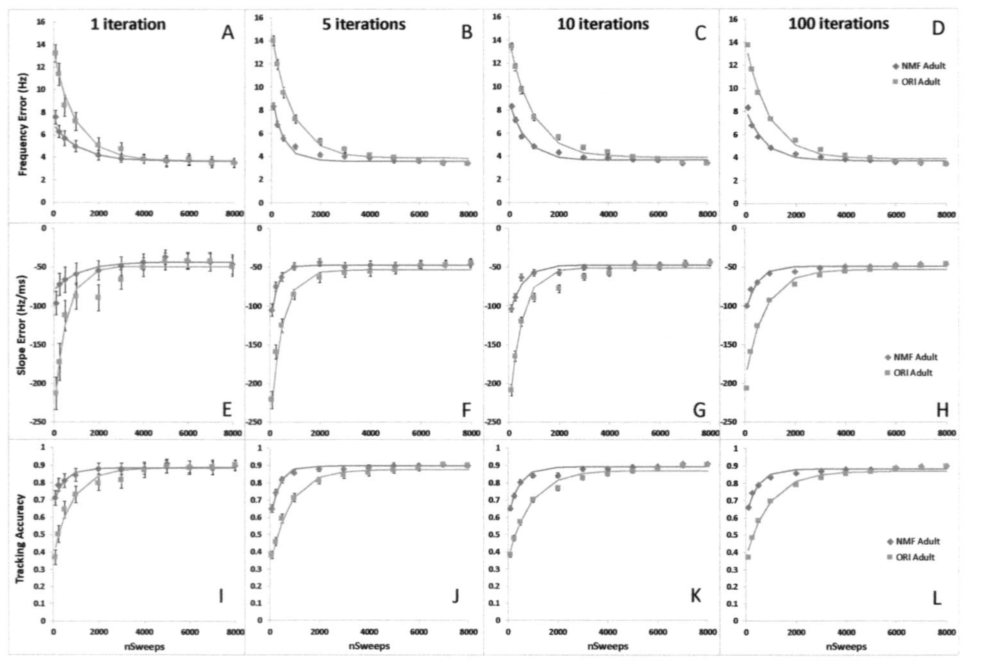

1 Iteration.

For FE, statistical results of the two-way ANOVA at 1 iteration demonstrated a significant difference for the nSweeps condition ($F = 52.579$, $p < 0.001$, $\eta_p^2 = 0.645$) and NMF condition ($F = 60.243$, $p < 0.001$, $\eta_p^2 = 0.675$) factors. as well as the interaction between the two factors ($F = 20.826$, $p < 0.001$, $\eta_p^2 = 0.418$). For the NMF condition factor that reached a significance, FE values were significantly smaller (mean difference $= 1.590 \pm 0.205$ Hz) after the NMF algorithm was applied. Furthermore, the results of all pair-wise comparisons across the 11 nSweeps conditions are summarized in Appendix C.

For SE, statistical results of the two-way ANOVA at 1 iteration demonstrated a significant difference for the nSweeps condition ($F = 18.872$, $p < 0.001$, $\eta_p^2 = 0.394$) and NMF condition ($F = 51.838$, $p < 0.001$, $\eta_p^2 = 0.641$) factors, as well as the interaction between the two factors ($F = 17.194$, $p < 0.001$, $\eta_p^2 = 0.372$). For the NMF condition factor that reached a significance, SE values were significantly larger (mean difference $= 32.878 \pm 4.566$ Hz/ms) after the NMF algorithm was applied. Finally, the results of all pair-wise comparisons across the 11 nSweeps conditions are also summarized in Appendix C.

In regard to the TA feature, statistical results of the two-way ANOVA at 1 iteration demonstrated a significant difference or the nSweeps condition ($F = 27.502$, $p < 0.001$, $\eta_p^2 = 0.487$) and NMF condition ($F = 67.908$, $p < 0.001$, $\eta_p^2 = 0.701$) factors, as well as the interaction between the two factors ($F = 19.312$, $p < 0.001$, $\eta_p^2 = 0.400$). For the NMF condition factor that reached a significance, TA values were significantly larger (mean difference $= 0.098 \pm 0.012$) after the NMF algorithm was applied. For clarity, results of all pair-wise comparisons across the 11 nSweeps conditions are also summarized in Appendix C.

5 Iterations.

For FE, statistical results of the two-way ANOVA at 5 iterations demonstrated a significant difference for the nSweeps condition ($F = 110.606$, $p < 0.001$, $\eta_p^2 = 0.792$) and NMF condition ($F = 77.450$, $p < 0.001$, $\eta_p^2 = 0.728$) factors, as well as the interaction between the two factors ($F = 51.066$, $p < 0.001$, $\eta_p^2 = 0.638$). For the NMF condition factor that reached a significance, FE values were significantly smaller (mean difference $= -1.769 \pm 0.201$ Hz) after the NMF algorithm was applied. Finally, the results of all pair-wise comparisons across the 11 nSweeps conditions are summarized in Appendix D.

In regard to the SE feature, statistical results of the two-way ANOVA at 5 iteration demonstrated a significant difference or the nSweeps condition ($F = 34.369$, $p < 0.001$, $\eta_p^2 = 0.542$) and NMF condition ($F = 46.979$, $p < 0.001$, $\eta_p^2 = 0.618$) factors, as well as the interaction between the two factors ($F = 52.872$, $p < 0.001$, $\eta_p^2 = 0.646$). For the NMF condition factor that reached a significance, SE values were significantly larger (mean difference $= 31.003 \pm 4.523$ Hz/ms) after the NMF algorithm was applied. For clarity, results of all pair-wise comparisons across the 11 nSweeps conditions are also summarized in Appendix D.

For TA with 5 iterations, the two-way ANOVA was significant for the nSweeps condition ($F = 73.993$, $p < 0.001$, $\eta_p^2 = 0.718$) and NMF condition ($F = 70.377$, $p < 0.001$, $\eta_p^2 = 0.708$) factors, as well as the interaction between the two factors ($F = 26.116$, $p < 0.001$, $\eta_p^2 = 0.474$). For the NMF condition factor, TA values were significantly larger (mean difference $= 0.100 \pm 0.012$) after the NMF algorithm was applied. Furthermore, the results of all pair-wise comparisons across the 11 nSweeps condition factor are also summarized in Appendix D.

10 Iterations.

For FE, statistical results of the two-way ANOVA at 10 iterations demonstrated a significant difference for the nSweeps condition ($F = 128.275$, $p < 0.001$, $\eta_p^2 = 0.816$) and NMF condition ($F = 101.659$, $p < 0.001$, $\eta_p^2 = 0.778$) factors, as well as the interaction between the two factors ($F = 50.340$, $p < 0.001$, $\eta_p^2 = 0.634$). For the NMF condition factor, FE values were significantly smaller (mean difference $= -1.764 \pm 0.174$ Hz) after the NMF algorithm was applied. Furthermore, the results of all pair-wise comparisons across the 11 nSweeps conditions are summarized in Appendix E.

For SE, statistical results of the two-way ANOVA at 10 iterations demonstrated a significant difference for the nSweeps condition ($F = 54.929$, $p < 0.001$, $\eta_p^2 = 0.654$) and NMF condition ($F = 48.045$, $p < 0.001$, $\eta_p^2 = 0.624$) factors, as well as the interaction between the two factors ($F = 56.781$, $p < 0.001$, $\eta_p^2 = 0.662$). For the NMF condition factor, SE values were significantly larger (mean difference $= 29.205 \pm 4.213$ Hz/ms) after the NMF algorithm was applied. Finally, the results of all pair-wise comparisons across the 11 nSweeps conditions are also summarized in Appendix E.

For TA, statistical results of the two-way ANOVA at 10 iterations demonstrated a significant difference for the nSweeps condition ($F = 84.239$, $p < 0.001$, $\eta_p^2 = 0.744$) and NMF condition ($F = 90.882$, $p < 0.001$, $\eta_p^2 = 0.624$) factors, as well as the interaction between the two factors ($F = 33.018$, $p < 0.001$, $\eta_p^2 = 0.532$). For the NMF condition factor, TA values were significantly larger (mean difference = 0.099 ± 0.010) after the NMF algorithm was applied. For clarity, results of all pair-wise comparisons across the 11 nSweeps conditions are also summarized in Appendix E.

100 Iterations.

For FE, statistical results of the two-way ANOVA at 100 iterations demonstrated a significant difference in the FE for the nSweeps condition ($F = 162.902$, $p < 0.001$, $\eta_p^2 = 0.849$) and NMF condition ($F = 99.748$, $p < 0.001$, $\eta_p^2 = 0.775$) factors, as well as the interaction between the two factors ($F = 80.548$, $p < 0.001$, $\eta_p^2 = 0.735$). For the NMF condition factor, FE values were significantly smaller (mean difference = -1.721 ± 0.172 Hz) after the NMF algorithm was applied. Furthermore, the results of all pair-wise comparisons across the 11 nSweeps conditions are summarized in Appendix F.

For SE, statistical results of the two-way ANOVA at 100 iterations demonstrated a significant difference for the nSweeps condition ($F = 72.504$, $p < 0.001$, $\eta_p^2 = 0.714$) and NMF condition ($F = 70.627$, $p < 0.001$, $\eta_p^2 = 0.709$) factors, as well as the interaction between the two factors ($F = 102.897$, $p < 0.001$, $\eta_p^2 = 0.780$). For the NMF condition factor, SE values were significantly larger (mean difference = 29.206 ± 3.475 Hz/ms) after the NMF algorithm was applied. Finally, the results of all pair-wise comparisons across the 11 nSweeps conditions are also summarized in Appendix F.

For TA, statistical results of the two-way ANOVA at 100 iterations demonstrated a significant difference for the nSweeps condition ($F = 105.508$, $p < 0.001$, $\eta_p^2 = 0.784$) and NMF condition ($F = 89.324$, $p < 0.001$, $\eta_p^2 = 0.755$) factors, as well as the interaction between the two factors ($F = 59.420$, $p < 0.001$, $\eta_p^2 = 0.672$). For the NMF condition factor, TA values were significantly larger (mean difference = 0.096 ± 0.010) after the NMF algorithm was applied. For clarity, results of all pair-wise comparisons across the 11 nSweeps conditions are also summarized in Appendix F.

Evaluation of the NMF Performance for Adults (Aim 2)

To reiterate, the purpose of Aim 2 was to present a novel approach to extract the "hidden" portion of neural responses, in addition to the "apparent" portion of neural responses.

Figure 9 depicts the results of hidden components extracted from the adult recordings. Panel A displayed the hFE results. For 1 iteration, the hidden component was smallest at 100 sweeps (-5.634 ± 0.801 Hz). With an increasing number of sweeps, hFE values began to increase. From 4000 sweeps on, the hidden component began to taper out and become very close to zero Hz. Results of this investigation demonstrated that the NMF algorithm was able to uncover a substantial amount of the hFE component. This phenomenon was particularly apparent when 100-4000 sweeps were included. When viewing data from 5, 10, and 100 iterations (Panels B, C, & D), results were consistent with those observed at 1 iteration.

Panel E displayed the hSE results. For 1 iteration, the hidden component was largest at 100 sweeps (117.310 ± 19.600 Hz/ms). With an increasing number of sweeps, the hSE began to decrease and became very closer to zero Hz/ms. From about 4000 sweeps, the hSE component plateaued at about 6.891 ± 6.078 Hz/ms. Results from this investigation indicated that the NMF algorithm was able to reveal a substantial amount of hSE values at 100-4000 sweeps. Please note that these hSE values were previously unavailable until the NMF algorithm was applied to each recording. When viewing data from 5, 10, and 100 iterations (F, G, & H), results were consistent with those observed at 1 iteration.

Results of the hTA feature can be viewed in Panel I. For 1 iteration, the hidden component was largest at 100 sweeps (0.340 ± 0.049). With an increasing number of sweeps, the hTA values began to decrease. From 4000 sweeps, the hidden component became very close to zero. In other words, a substantial amount of hTA was uncovered when the NMF algorithm was applied at 100-4000 sweeps. When viewing data from 5, 10, and 100 iterations (J, K, & L), results were consistent with those observed at 1 iteration.

Figure 9

Graphical presentations display the raw data extracts for the adult participants hidden component at 1, 5, 10 and 100 iterations for Hidden Frequency Error (first row), Hidden Slope Error (second row), and Hidden Tracking Accuracy (third row).

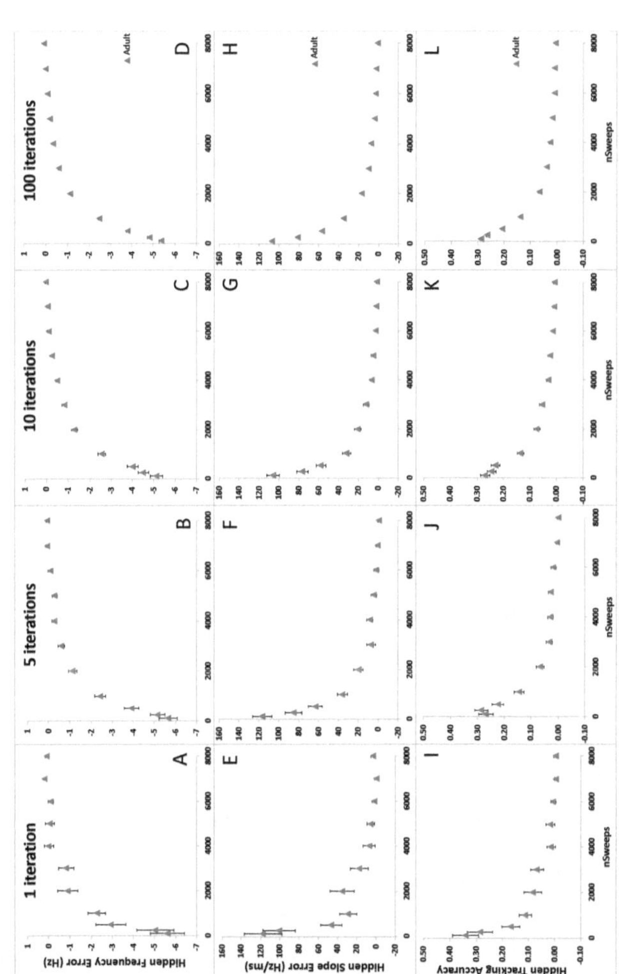

Linear Regression

Results of the linear regression analysis for the adult hidden component can be seen in Figure 10. Panel A depicts the results of the linear regression with the nSweeps condition on the x-axis and hFE on the y-axis, Panel E depicts the results of the linear regression with the nSweeps condition on the x-axis and hSE on the y-axis, and Panel I depicts the results of the linear regression with the nSweeps condition on the x-axis and hTA on the y-axis. Each column in the figure displayed 1, 5, 10, and 100 iterations, respectively. The resultant linear regression equations can be viewed below.

1 Iteration.

$$hFE \text{ Linear Regression: } y = 0.0006x + 3.647 \tag{55}$$

$$hSE \text{ Linear Regression: } y = -0.0113x + 70.863 \tag{56}$$

$$hTA \text{ Linear Regression: } y = -4E\text{-}05x + 0.2163 \tag{57}$$

Slope estimates for Equation 55 depicts the hFE results. The hFE slope confirmed an incremental trend for hFE as nSweeps increased. The slope for hSE (Equation 56) indicated a decremental trend for hSE as nSweeps increased. Finally, the hTA slope (Equation 57) indicated a decreasing trend in hTA as nSweeps increased.

5 Iterations.

$$hFE \text{ Linear Regression: } y = 0.0007x - 3.9501 \tag{58}$$

$$hSE \text{ Linear Regression: } y = -0.0114x + 69.198 \tag{59}$$

$$hTA \text{ Linear Regression: } y = -3E\text{-}05x + 0.2129 \tag{60}$$

When viewing the 5 iteration equations, the values and trends remained similar to the 1 iteration results.

10 Iterations.

hFE Linear Regression: y = 0.0006x + 3.647 (61)

hSE Linear Regression: y = -0.0101x + 63.132 (62)

hTA Linear Regression: y = -3E-05x + 0.2056 (63)

When viewing the 10 iteration equations, the values and trends remained similar to the 1 and 5 iteration results.

100 Iterations.

hFE Linear Regression: y = 0.0006x - 3.797 (64)

hSE Linear Regression: y = -0.0104x + 64.092 (65)

hTA Linear Regression: y = -3E-05x + 0.2065 (66)

When viewing the 100 iteration equations, the values and trends remained similar to the 1, 5, and 10 iteration results.

Figure 10

Graphical depictions of the linear regressions performed on the hidden component for the adult participants at 1, 5, 10 and 100 iterations for Hidden Frequency Error (first row), Hidden Slope Error (second row), and Hidden Tracking Accuracy (third row).

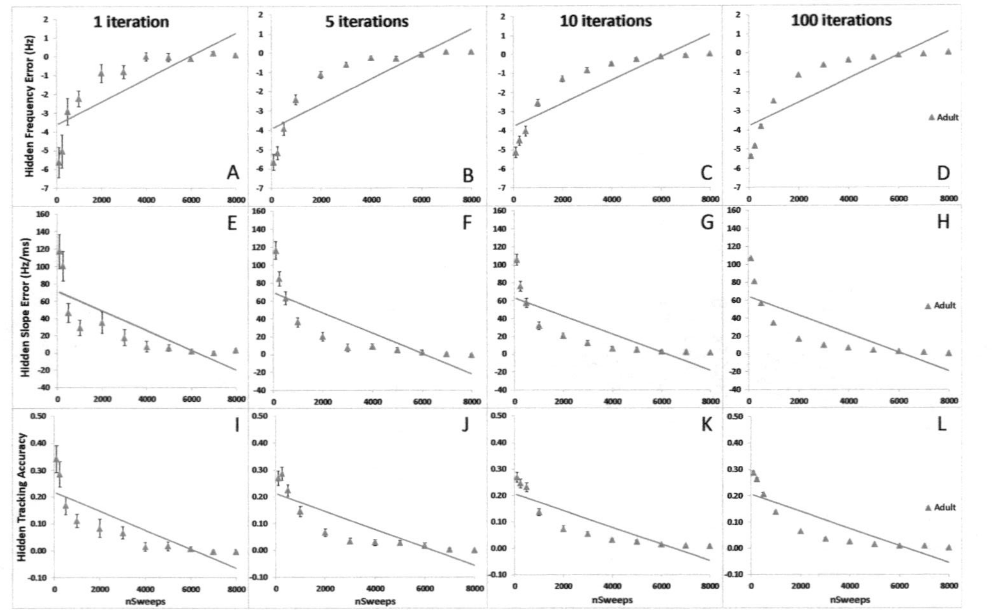

Exponential Modeling

Results of exponential modeling for the adult hidden component can be seen in Figure 11. Similar to the previous figures, Panel A depicts the results of hFE, Panel E depicts the results of hSE, and Panel I depicts the results of hTA. Each column in the figure 1, 5, 10, and 100 iterations, respectively. The derived equations can be viewed below. A tabular depiction of the Tau, A_{noise}, and A_{as} values can be viewed in Appendix G.

1 Iteration.

Adult Exponential Modeling for hFE: $A(n) = -0.015\left(1 - e^{-\frac{n}{972.560}}\right) + 5.634$ (67)

Adult Exponential Modeling for hSE: $A(n) = 117.310\left(e^{-\frac{n}{577.063}}\right) - 5.354$ (68)

Adult Exponential Modeling for hTA: $A(n) = 0.012\left(e^{-\frac{n}{751.918}}\right) - 0.340$ (69)

The exponential modeling Equations 67, 68, and 69 indicated the overall trends for hFE, hSE, and hTA, respectively. The asymptotic amplitude (A_{AS}) for the hFE trend was -0.015 Hz, the A_{AS} value for the hSE trend is 5.354 Hz/ms, and for hTA the A_{AS} value was 0.340. The τ values of the fitted curves for the hFE, hSE, and hTA trends were 972, 577, and 751 sweeps, respectively. The noise amplitude (A_{noise}) values for hFE, hSE, and hTA trends were -5.634 Hz, 117.310 Hz/ms and 0.012, respectively. These values aided in the depiction of the hidden components' overall trends and plateaus within the exponential model.

5 Iterations.

Adult Exponential Modeling for hFE: $A(n) = -0.064\left(1 - e^{-\frac{n}{1153.987}}\right) + 5.666$ (70)

Adult Exponential Modeling for hSE: $A(n) = 102.178\left(e^{-\frac{n}{795.862}}\right) - 4.224$ (71)

Adult Exponential Modeling for hTA: $A(n) = 0.012\left(e^{-\frac{n}{1106.627}}\right) - 0.317$ (72)

When viewing the 5 iteration equations, the values remained similar to the 1 iteration results.

10 Iterations.

Adult Exponential Modeling for hFE: $A(n) = -0.057\left(1 - e^{-\frac{n}{1352.667}}\right) + 5.147$ (73)

Adult Exponential Modeling for hSE: $A(n) = 95.312\left(e^{-\frac{n}{742.458}}\right) - 4.849$ (74)

Adult Exponential Modeling for hTA: $A(n) = 0.009\left(e^{-\frac{n}{1577.897}}\right) - 0.269$ (75)

When viewing the 10 iteration equations, the values remained similar to the 1 and 5 iteration results.

100 Iterations.

Adult Exponential Modeling for hFE: $A(n) = -0.073\left(1 - e^{-\frac{n}{1194.074}}\right) + 5.368$ (76)

Adult Exponential Modeling for hSE: $A(n) = 96.525\left(e^{-\frac{n}{823.570}}\right) - 3.554$ (77)

Adult Exponential Modeling for hTA: $A(n) = 0.007\left(e^{-\frac{n}{1212.950}}\right) - 0.289$ (78)

When viewing the 100 iteration equations, the values remained similar to the 1, 5, and 10 iteration results.

Figure 11

Graphical displays of the exponential modeling performed on the adult hidden component at 1, 5, 10 and 100 iterations for Hidden Frequency Error (first row), Hidden Slope Error (second row), and Hidden Tracking Accuracy (third row).

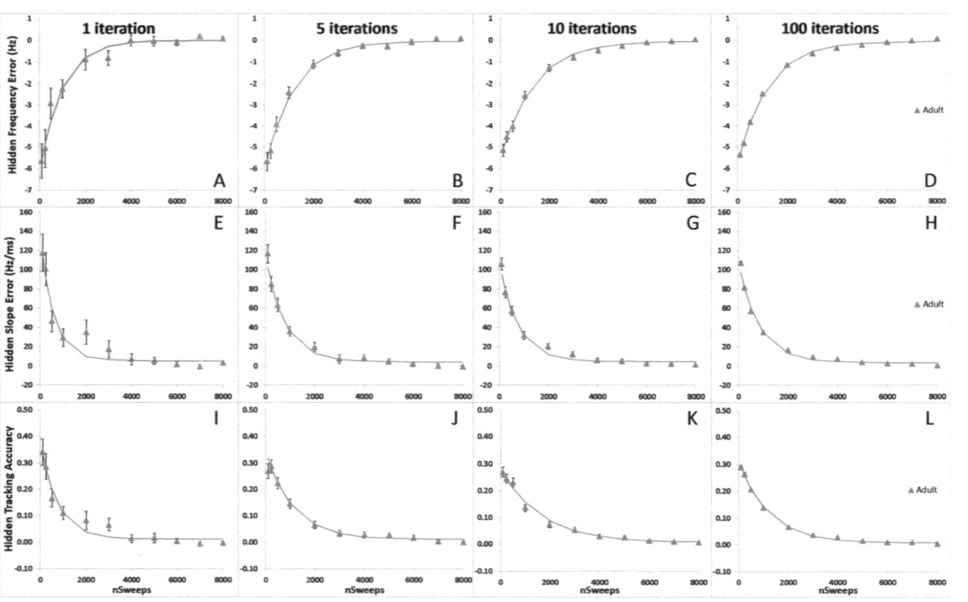

One-way ANOVA

1 Iteration.

For hFE, statistical results of the one-way ANOVA at 1 iteration demonstrated a significant difference for the nSweeps condition (F = 17.583, $p < 0.001$, $\eta_p^2 = 0.403$) factor. Results of all pair-wise comparisons are summarized in Appendix C.

For hSE, statistical results of the one-way ANOVA at 1 iteration demonstrated a significant difference for the nSweeps condition (F = 13.025, $p < 0.001$, $\eta_p^2 = 0.334$) factor. Results of all pair-wise comparisons are also summarized in Appendix C.

For hTA, statistical results of the one-way ANOVA at 1 iteration demonstrated a significant difference for the nSweeps condition (F = 14.712, $p < 0.001$, $\eta_p^2 = 0.361$) factor. Results of all pair-wise comparisons are also summarized in Appendix C.

5 Iterations.

For hFE, statistical results of the one-way ANOVA at 5 iterations demonstrated a significant difference for the nSweeps condition (F = 47.715, $p < 0.001$, $\eta_p^2 = 0.630$) factor. Results of all pair-wise comparisons are summarized in Appendix D.

For hSE, statistical results of the one-way ANOVA at 5 iterations demonstrated a significant difference for the nSweeps condition (F = 49.201, $p < 0.001$, $\eta_p^2 = 0.637$) factor. Results of all pair-wise comparisons are also summarized in Appendix D.

For hTA, statistical results of the one-way ANOVA at 5 iterations demonstrated a significant difference for the nSweeps condition (F = 24.176, $p < 0.001$, $\eta_p^2 = 0.463$) factor. Results of all pair-wise comparisons are also summarized in Appendix D.

10 Iterations.

For hFE, statistical results of the one-way ANOVA at 10 iterations demonstrated a significant difference for the nSweeps condition ($F = 38.485$, $p < 0.001$, $\eta_p^2 = 0.606$) factor. Results of all pair-wise comparisons are summarized in Appendix E.

For hSE, statistical results of the one-way ANOVA at 10 iterations demonstrated a significant difference for the nSweeps condition ($F = 46.067$, $p < 0.001$, $\eta_p^2 = 0.648$) factor. Results of all pair-wise comparisons are also summarized in Appendix E.

For hTA, statistical results of the one-way ANOVA at 10 iterations demonstrated a significant difference for the nSweeps condition ($F = 24.636$, $p < 0.001$, $\eta_p^2 = 0.496$) factor. Results of all pair-wise comparisons are also summarized in Appendix E.

100 Iterations.

For hFE, statistical results of the one-way ANOVA at 100 iterations demonstrated a significant difference for the nSweeps condition ($F = 70.212$, $p < 0.001$, $\eta_p^2 = 0.722$) factor. Results of all pair-wise comparisons are summarized in Appendix F.

For hSE, statistical results of the one-way ANOVA at 100 iterations demonstrated a significant difference for the nSweeps condition ($F = 90.495$, $p < 0.001$, $\eta_p^2 = 0.770$) factor. Results of all pair-wise comparisons are also summarized in Appendix F.

For hTA, statistical results of the one-way ANOVA at 100 iterations demonstrated a significant difference for the nSweeps condition ($F = 50.999$, $p < 0.001$, $\eta_p^2 = 0.654$) factor. Results of all pair-wise comparisons are also summarized in Appendix F.

Application of the NMF Algorithm for Newborns (Aim 3a)

The purpose of Aim 3a was to systematically manipulate NMF parameters and investigate their performances of noise separation from FFRs in neonates.

The neonatal ORI and NMF results can be seen in Figure 12. Panel A displayed the FE results. For the ORI condition at 1 iteration, when the conventional signal averaging was utilized, the FFR was 14.398 ± 1.185 Hz at 100 sweeps. With an increasing number of sweeps, FE values decreased. Roughly 7000 sweeps were needed to see FE value stability. When visualizing results of the NMF condition, similar trends were observed. However, differences were also observed in the FE values between the ORI and NMF conditions (9.202 ± 0.627 Hz for the NMF condition and 14.398 ± 1.185 Hz for the ORI condition). Another difference was that roughly 3000-6000 sweeps were required to see a FE plateau in the NMF condition (instead of about 7000 sweeps in the ORI condition). When comparing the two results (ORI & NMF), it was apparent that the NMF condition required fewer sweeps in order to obtain comparable results than the ORI condition. There was a visual improvement in the FE values when the NMF algorithm was implemented. When viewing data from 5, 10, and 100 iterations (B, C, & D), results remained consistent with the 1 iteration findings.

Panel E displayed the results for the SE feature. For the ORI condition at 1 iteration, SE features were largest (-265.120 ± 20.303 Hz/ms) at 100 sweeps. As nSweeps increased, SE values increased and became closer to zero Hz/ms. Approximately 7000-8000 sweeps were needed to observe a SE plateau. For the NMF condition, the FFR was -142.206 ± 15.207 Hz/ms at 100 sweeps. Roughly 1000-4000 sweeps were required to visualize an SE plateau when the NMF algorithm was implemented. SE values showed a visual improvement in the NMF condition. Similarly, the data from 5, 10, and 100 iterations (F, G, & H) were in line with the results of 1 iteration.

The results of the TA feature can be viewed in Panel I. For the ORI condition at 1 iteration, at 100 sweeps the FFR was 0.325 ± 0.045. With an increasing number of sweeps, TA values increased and plateaued at approximately 8000 sweeps. When looking at the NMF condition, similar trends were observed. However, there were also differences among the TA values between the ORI and NMF conditions (0.492 ± 0.052 for the NMF condition and 0.325 ± 0.045 for the ORI condition). Another difference was that the TA plateau occurred roughly between 4000-6000 sweeps in the NMF condition (instead of about 8000 sweeps in the ORI condition). Meaning that, the NMF condition required fewer sweeps in order to obtain comparable results than the ORI condition. When comparing data from 5, 10, and 100 iterations (J, K, & L), results remained consistent with those observed at 1 iteration.

Figure 12

Graphical presentations of the raw data extracts for the ORI and NMF conditions for the newborn participants at 1, 5, 10 and 100 iterations for Frequency Error (first row), Slope Error (second row), and Tracking Accuracy (third row).

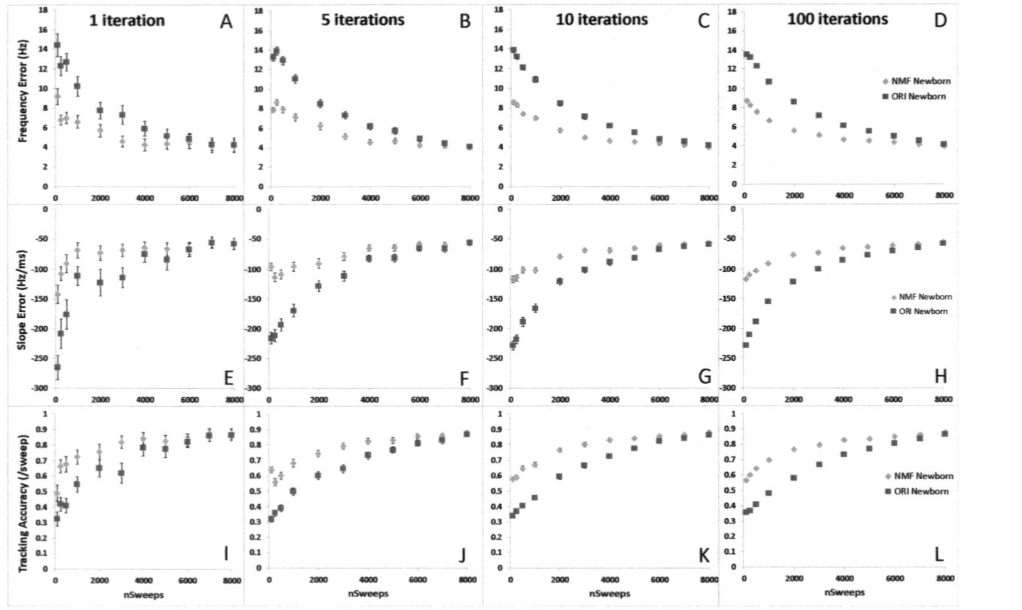

Linear Regression

Linear regressions conducted for the ORI and NMF newborn data are displayed in Figure 13. Formatting and layout for Figure 13 was consistent with those shown in the previous figures. Equations for all linear regression results can be viewed below.

1 Iteration.

ORI Linear Regression for FE $y = -0.012x + 12.12$ (79)

NMF Linear Regression for FE: $y = -0.001x + 7.21$ (80)

ORI Linear Regression for SE: $y = 0.0202x - 189.66$ (81)

NMF Linear Regression for SE: $y = 0.0067x - 100.84$ (82)

ORI Linear Regression for TA: $y = 6E\text{-}05x + 0.4281$ (83)

NMF Linear Regression for TA: $y = 3E\text{-}05x + 0.6463$ (84)

For FE, slope estimates for Equation 79 and 80 displayed differing values. The ORI slope was steeper than that of the NMF condition. This meant that, for the ORI condition, FE values decreased at a faster rate when compared to those observed in the NMF condition. In other words, the decrement for the ORI FE took a smaller number of nSweeps to obtain a similar performance than the decrement in the NMF condition. When viewing the ORI and NMF intercepts, values for the NMF condition indicated a lower noise floor for the FE trend than those observed in the ORI condition.

When looking at SE (Equations 81 and 82), there were differing slope estimates. The ORI slope was steeper than the NMF slope. This means that, for the ORI condition, SE values increased at a faster rate compared to those in the NMF condition. In other words, the increment for the ORI SE took a smaller number of nSweeps to obtain a similar performance to the NMF SE. When viewing the differing SE intercepts, values for the NMF condition indicated a lower noise floor for the SE trend than those observed in the ORI condition.

Slope estimates for Equation 83 and 84 displayed differing values. For TA, the NMF slope was steeper than the ORI slope. Simply put, the trend for the TA values in the NMF condition took a smaller number of sweeps to increase when compared to the ORI condition. In other words, the increment for the NMF condition took a smaller number of nSweeps to obtain a similar performance when compared to the ORI condition. When viewing the differing intercepts, values for the NMF condition could indicate a lower noise floor for the TA trend.

5 Iterations.

$$\text{ORI Linear Regression for FE } y = -0.0012x + 12.564 \tag{85}$$

$$\text{NMF Linear Regression for FE: } y = -0.0006x + 7.7629 \tag{86}$$

$$\text{ORI Linear Regression for SE: } y = 0.0205x - 194.28 \tag{87}$$

$$\text{NMF Linear Regression for SE: } y = 0.007x - 104.15 \tag{88}$$

$$\text{ORI Linear Regression for TA: } y = 7E\text{-}05x + 0.3945 \tag{89}$$

$$\text{NMF Linear Regression for TA: } y = 4E\text{-}05x + 0.6276 \tag{90}$$

When viewing the 5 iteration equations, the values and trends remained similar to the 1 iteration results.

10 Iterations.

$$\text{ORI Linear Regression for FE } y = -0.0012x + 12.333 \tag{91}$$

$$\text{NMF Linear Regression for FE: } y = -0.0005x + 7.6348 \tag{92}$$

$$\text{ORI Linear Regression for SE: } y = 0.0206x - 194.72 \tag{93}$$

$$\text{NMF Linear Regression for SE: } y = 0.0075x - 106.53 \tag{94}$$

ORI Linear Regression for TA: y = 7E-05x + 0.3969 (95)

NMF Linear Regression for TA: y = 4E-05x + 0.6325 (96)

When viewing the 10 iteration equations, the values and trends remained similar to the 1 and 5 iteration results.

100 Iterations.

ORI Linear Regression for FE y = -0.0012x + 12.274 (97)

NMF Linear Regression for FE: y = -0.0006x + 7.624 (98)

ORI Linear Regression for SE: y = 0.02x - 190.73 (99)

NMF Linear Regression for SE: y = 0.01x - 103.46 (100)

ORI Linear Regression for TA: y = 7E-05x + 0.4047 (101)

NMF Linear Regression for TA: y = 4E-05x + 0.6358 (102)

When viewing the 100 iteration equations, the values and trends remained similar to the 1, 5, and 10 iteration results.

Figure 13

Graphical presentations of the linear regressions performed on the ORI and NMF conditions for the newborn participants at 1, 5, 10 and 100 iterations for Frequency Error (first row), Slope Error (second row), and Tracking Accuracy (third row).

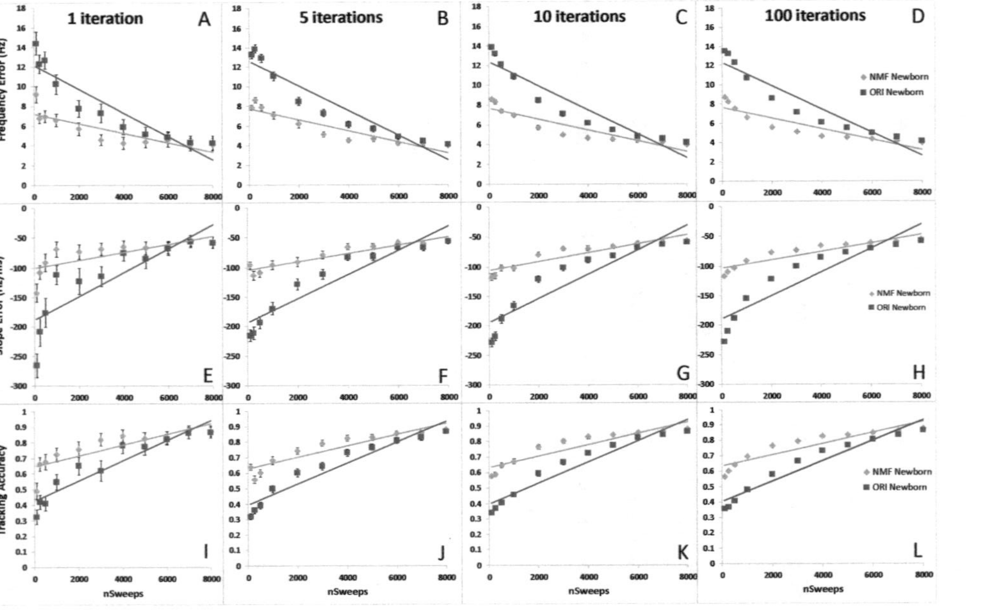

Exponential Modeling

Exponential modeling was conducted on data obtained in the newborn ORI and NMF conditions. Results can be seen in Figure 14 (ORI and NMF). Formatting in these two figures was consistent with previous figures. Panel A depicts the results of FE, Panel E depicts the results of SE, and Panel I depicts the results of TA. Exponential modeling equations are listed below by iteration. A tabular depiction of the Tau, A_{noise}, and A_{as} values can be viewed in Appendix G.

1 Iteration.

$$\text{ORI Exponential Modeling for FE: } A(n) = 14.221\left(e^{-\frac{n}{2327.450}}\right) - 4.254 \tag{103}$$

$$\text{NMF Exponential Modeling for FE: } A(n) = 7.783\left(e^{-\frac{n}{2050.346}}\right) - 4.104 \tag{104}$$

$$\text{ORI Exponential Modeling for SE: } A(n) = -74.608\left(1 - e^{-\frac{n}{634.157}}\right) + 265.110 \tag{105}$$

$$\text{NMF Exponential Modeling for SE: } A(n) = -65.435\left(1 - e^{-\frac{n}{282.656}}\right) + 142.206 \tag{106}$$

$$\text{ORI Exponential Modeling for TA: } A(n) = 0.863\left(1 - e^{-\frac{n}{2300.823}}\right) - 0.333 \tag{107}$$

$$\text{NMF Exponential Modeling for TA: } A(n) = 0.873\left(1 - e^{-\frac{n}{2047.339}}\right) - 0.634 \tag{108}$$

For FE, the exponential modeling Equations 103 and 104 displayed differing values. The noise amplitude (A_{noise}) values for the ORI and NMF conditions were 14.221 and 7.783 Hz, respectively. These values indicated that the A_{noise} in the NMF condition was 45% ([14.221 - 7.783] /14.221 = 45%) lower than that of the ORI condition. This indicated that a 45% increment in the A_{noise} estimates of NMF FE values, demonstrating a better performance in the overall FE trends. The τ values of the fitted curves for the FE trends obtained in the ORI and NMF conditions were 2327 and 2050 sweeps, respectively. It was important to note that the NMF produced a smaller τ value than the

ORI. This finding indicated that the NMF condition showed a faster improvement in frequency-encoding sharpness (i.e., less Frequency Error) with increasing number of sweeps when compared to the NMF condition. The asymptotic amplitude (A_{As}) for the ORI and NMF conditions were 4.254 Hz and 4.104 Hz, respectively. Similar FE plateau values indicated that the NMF algorithm had little to no influence on the FE plateau.

For SE, the exponential modeling Equations 105 and 106 displayed differing values. The asymptotic amplitude (A_{As}) for the ORI and NMF conditions were -74.608 Hz/ms and -65.435 Hz/ms, respectively. Similar SE plateau values indicated that the NMF algorithm had little to no influence on the SE plateau. The τ values of the fitted curves for the SE trends obtained in the ORI and NMF participants were 634 and 282 sweeps, respectively. It was important to note that the NMF condition produced a smaller τ value than the ORI condition. This finding meant that the NMF condition showed a faster improvement in SE with increasing nSweeps compared to the ORI condition. The noise amplitude (A_{noise}) values for the ORI and NMF conditions are -265.110 and -142.206 Hz/ms, respectively. These values signified that the A_{noise} in the NMF condition was 46% ([-265.110 + 142.206] /-265.110= 46%) lower than that of the ORI condition. This indicated a 46% increment in the A_{noise} estimates of the NMF SE values, showing a better performance in the overall SE trends.

For TA, the exponential modeling Equations 107 and 108 displayed differing values. The asymptotic amplitude (A_{As}) for the ORI and NMF conditions were 0.863 and 0.873, respectively. Similar TA plateau values indicated that the NMF algorithm has little to no influence on the TA plateau. The τ values of the fitted curves for TA were obtained in the ORI and NMF participants were 2300 and 2047 sweeps, respectively. It was important to note that the NMF condition produced a smaller τ value than the ORI condition. This finding indicated that the NMF condition showed a faster improvement in pitch encoding (i.e., better tracking accuracy) with increasing number of sweeps compared to the ORI. The A_{noise} values for the ORI and NMF conditions were 0.333 and 0.634, respectively. These values indicated that the A_{noise} in the ORI condition was 48% ([0.634- 0.333] /0.633 = 48%) lower than that of the NMF condition. This indicated a 48% increment in the A_{noise} estimates of the OR SE values, showing a better performance in overall TA trends.

5 Iterations.

ORI Exponential Modeling for FE: $A(n) = 13.763\left(e^{-\frac{n}{2939.978}}\right) - 4.160$ (109)

NMF Exponential Modeling for FE: $A(n) = 8.681\left(e^{-\frac{n}{2241.782}}\right) - 4.030$ (110)

ORI Exponential Modeling for SE: $A(n) = -56.154\left(1 - e^{-\frac{n}{2799.854}}\right) + 216.340$ (111)

NMF Exponential Modeling for SE: $A(n) = -55.460\left(1 - e^{-\frac{n}{3446.356}}\right) + 116.323$ (112)

ORI Exponential Modeling for TA: $A(n) = 0.851\left(1 - e^{-\frac{n}{3221.558}}\right) - 0.333$ (113)

NMF Exponential Modeling for TA: $A(n) = 0.871\left(1 - e^{-\frac{n}{2215.412}}\right) - 0.557$ (114)

When viewing the 5 iteration equations, the A_{Noise}, τ, and A_{AS} values for FE and TA remained similar to the 1 iteration results. For SE, the exponential modeling Equations 111 and 112 displayed differing values, when compared to 1 iteration. The SE τ values of the fitted curves for the ORI and NMF participants were 2799 and 3446 sweeps, respectively. It was important to note that the NMF condition produced a larger τ value than the ORI condition. This finding indicated that the ORI condition showed a faster improvement in SE with increasing number of sweeps compared to the NMF condition. The SE A_{Noise} and A_{AS} values at 5 iterations were consistent with the 1 iteration results.

10 Iterations.

$$\text{ORI Exponential Modeling for FE: } A(n) = 13.796\left(e^{-\frac{n}{2577.217}}\right) - 4.357 \tag{115}$$

$$\text{NMF Exponential Modeling for FE: } A(n) = 8.571\left(e^{-\frac{n}{1839.548}}\right) - 4.177 \tag{116}$$

$$\text{ORI Exponential Modeling for SE: } A(n) = -62.061\left(1 - e^{-\frac{n}{1958.125}}\right) + 227.979 \tag{117}$$

$$\text{NMF Exponential Modeling for SE: } A(n) = -57.654\left(1 - e^{-\frac{n}{2315.143}}\right) + 117.387 \tag{118}$$

$$\text{ORI Exponential Modeling for TA: } A(n) = 0.863\left(1 - e^{-\frac{n}{3661.855}}\right) - 0.347 \tag{119}$$

$$\text{NMF Exponential Modeling for TA: } A(n) = 0.867\left(1 - e^{-\frac{n}{2046.337}}\right) - 0.578 \tag{120}$$

When viewing the 10 iteration equations, the values remained similar to the 5 iteration results.

100 Iterations.

$$\text{ORI Exponential Modeling for FE: } A(n) = 13.498\left(e^{-\frac{n}{2722.601}}\right) - 4.297 \tag{121}$$

$$\text{NMF Exponential Modeling for FE: } A(n) = 8.840\left(e^{-\frac{n}{1598.838}}\right) - 4.280 \tag{122}$$

$$\text{ORI Exponential Modeling for SE: } A(n) = -65.550\left(1 - e^{-\frac{n}{1909.178}}\right) + 222.658 \tag{123}$$

$$\text{NMF Exponential Modeling for SE: } A(n) = -60.150\left(1 - e^{-\frac{n}{1636.003}}\right) + 115.059 \tag{124}$$

$$\text{ORI Exponential Modeling for TA: } A(n) = 0.855\left(1 - e^{-\frac{n}{3636.209}}\right) - 0.350 \tag{125}$$

$$\text{NMF Exponential Modeling for TA: } A(n) = 0.854\left(1 - e^{-\frac{n}{1662.349}}\right) - 0.576 \tag{126}$$

When viewing the 100 iteration equations, the values remained similar to the 1 iteration results.

Figure 14

Graphical displays of the exponential modeling performed on the newborn ORI and NMF data at 1, 5, 10 and 100 iterations for Frequency Error (first row), Slope Error (second row), and Tracking Accuracy (third row).

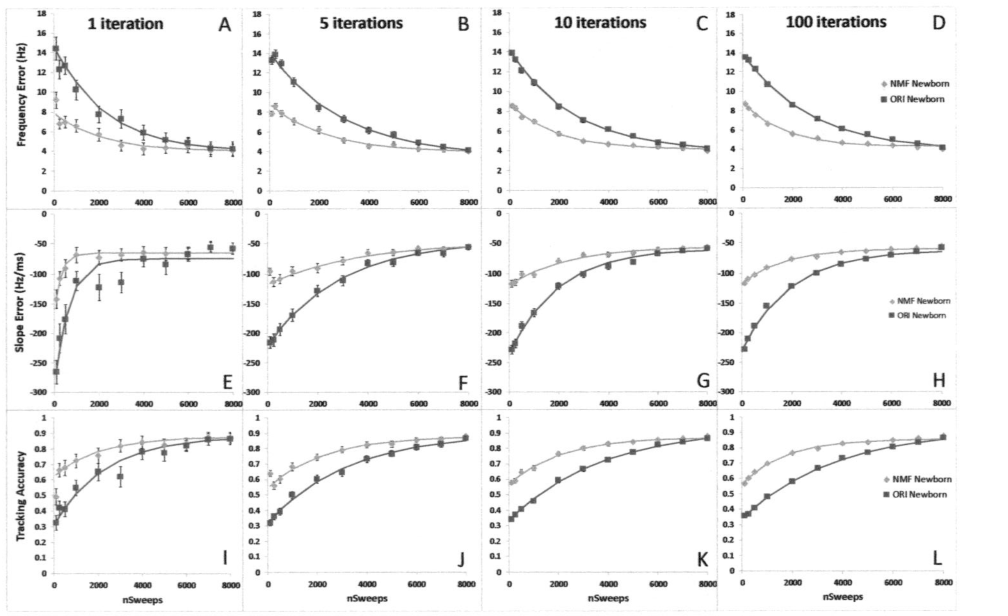

Two-way ANOVA

1 Iterations.

For FE, statistical results of the two-way ANOVA at 1 iteration demonstrated a significant difference for the nSweeps condition ($F = 33.673$, $p < 0.001$, $\eta_p^2 = 0.537$) and NMF condition ($F = 120.067$, $p < 0.001$, $\eta_p^2 = 0.805$) factors, as well as the interaction between the two factors ($F = 16.262$, $p< 0.001$, $\eta_p^2 = 0.359$). For the NMF condition factor that reached a significance, FE values were significantly smaller (mean difference $= -2.527 \pm 0.231$ Hz) after the NMF algorithm was applied. Finally, results of all pair-wise comparisons across the 11 nSweeps conditions are summarized in Appendix C.

For SE, statistical results of the two-way ANOVA at 1 iteration demonstrated a significant difference for the nSweeps condition ($F = 17.832$, $p< 0.001$, $\eta_p^2 = 0.381$) and NMF condition ($F = 68.164$, $p < 0.001$, $\eta_p^2 = 0.702$) factors, as well as the interaction between the two factors ($F = 9.640$, $p < 0.001$, $\eta_p^2 = 0.249$). For the NMF condition factor that reached a significance, SE values were significantly larger (mean difference $= 43.516 \pm 5.271$ Hz/ms) after the NMF algorithm was applied. For clarity, results of all pair-wise comparisons across the 11 nSweeps conditions are also summarized in Appendix C.

For TA, statistical results of the two-way ANOVA at 1 iteration demonstrated a significant difference for the nSweeps condition ($F = 26.811$, $p < 0.001$, $\eta_p^2 = 0.480$) and NMF condition ($F = 43.252$, $p < 0.001$, $\eta_p^2 = 0.599$) factors, as well as the interaction between the two factors ($F = 5.694$, $p < 0.001$, $\eta_p^2 = 0.164$).For the NMF condition factor that reached a significance, TA values were significantly larger (mean difference $= 0.118 \pm 0.018$) after the NMF algorithm was applied. Furthermore, results of all pair-wise comparisons across the 11 nSweeps conditions are also summarized in Appendix C.

5 Iterations.

For FE, statistical results of the two-way ANOVA at 5 iterations demonstrated a significant difference for the nSweeps condition ($F = 93.730$, $p < 0.001$, $\eta_p^2 = 0.764$) and NMF condition ($F = 252.428$, $p < 0.001$, $\eta_p^2 = 0.897$) factors, as well as the interaction between the two factors ($F = 45.003$, $p < 0.001$, $\eta_p^2 = 0.608$). For the NMF factor that reached a significance, FE values were significantly smaller (mean difference = -2.517 ± 0.158 Hz) after the NMF algorithm was applied. For clarity, results of all pair-wise comparisons across the 11 nSweeps conditions are summarized in Appendix D.

In regard to SE, statistical results of the two-way ANOVA at 5 iterations demonstrated a significant difference for the nSweeps condition ($F = 38.152$, $p < 0.001$, $\eta_p^2 = 0.568$) and NMF condition ($F = 138.862\ 979$, $p < 0.001$, $\eta_p^2 = 0.827$) factors, as well as the interaction between the two factors ($F = 45.339$, $p < 0.001$, $\eta_p^2 = 0.610$). For the NMF factor that reached a significance, SE values were significantly larger (mean difference = 45.117 ± 3.829 Hz/ms) after the NMF algorithm was applied. Furthermore, the results of all pair-wise comparisons across the 11 nSweeps conditions are also summarized in Appendix D.

For TA, statistical results of the two-way ANOVA at 5 iterations demonstrated a significant difference for the nSweeps condition ($F = 69.918$, $p < 0.001$, $\eta_p^2 = 0.707$) and NMF condition ($F = 115.882$, $p < 0.001$, $\eta_p^2 = 0.800$) factors, as well as the interaction between the two factors ($F = 13.510$, $p < 0.001$, $\eta_p^2 = 0.318$). For the NMF factor that reached a significance, TA values were significantly larger (mean difference = 0.131 ±

0.012) after the NMF algorithm was applied. Finally, the results of all pair-wise comparisons across the 11 nSweeps conditions are also summarized in Appendix D.

10 Iterations.

When looking at FE, statistical results of the two-way ANOVA at 10 iterations demonstrated a significant difference for the nSweeps condition ($F = 133.042$, $p < 0.001$, $\eta_p^2 = 0.821$) and NMF condition ($F = 252.337$, $p < 0.001$, $\eta_p^2 = 0.897$) factors, as well as the interaction between the two factors ($F = 78.958$, $p < 0.001$, $\eta_p^2 = 0.731$). For the NMF condition factor that reached a significance, FE values were significantly smaller (mean difference = -2.473 ± 0.156 Hz) after the NMF algorithm was applied. Finally, the results of all pair-wise comparisons across the 11 nSweeps conditions are summarized in Appendix E.

For SE, Statistical results of the two-way ANOVA at 10 iterations demonstrated a significant difference for the nSweeps condition ($F = 59.009$, $p < 0.001$, $\eta_p^2 = 0.670$) and NMF condition ($F = 120.229$, $p < 0.001$, $\eta_p^2 = 0.806$) factors, as well as the interaction between the two factors ($F = 57.245$, $p < 0.001$, $\eta_p^2 = 0.664$). For the NMF condition factor that reached a significance, SE values were significantly larger (mean difference = 44.299 ± 4.040 Hz/ms) after the NMF algorithm was applied. Furthermore, the results of all pair-wise comparisons across the 11 nSweeps conditions are also summarized in Appendix E.

In regard to TA, Statistical results of the two-way ANOVA at 10 iterations demonstrated a significant difference for the nSweeps condition ($F = 101.813$, $p < 0.001$, $\eta_p^2 = 0.778$) and NMF condition ($F = 145.019$, $p < 0.001$, $\eta_p^2 = 0.833$) factors, as well as the interaction between the two factors ($F = 25.627$, $p < 0.001$, $\eta_p^2 = 0.469$). For the NMF condition factor that reached a significance, TA values were significantly larger (mean difference $= 0.132 \pm 0.011$) after the NMF algorithm was applied. For clarity, results of all pair-wise comparisons across the 11 nSweeps conditions are also summarized in Appendix E.

100 Iterations.

For FE, statistical results of the two-way ANOVA at 100 iterations demonstrated a significant difference for the nSweeps condition ($F = 209.777$, $p < 0.001$, $\eta_p^2 = 0.879$) and NMF condition ($F = 261.627$, $p < 0.001$, $\eta_p^2 = 0.900$) factors, as well as the interaction between the two factors ($F = 124.408$, $p < 0.001$, $\eta_p^2 = 0.811$). For the NMF condition factor that reached a significance, FE values were significantly smaller (mean difference $= -2.474 \pm 0.153$ Hz) after the NMF algorithm was applied. Furthermore, the results of all pair-wise comparisons across the 11 nSweeps conditions are summarized in Appendix F.

For SE, statistical results of the two-way ANOVA at 100 iterations demonstrated a significant difference for the nSweeps condition ($F = 158.710$, $p < 0.001$, $\eta_p^2 = 0.846$) and NMF condition ($F = 152.811$, $p < 0.001$, $\eta_p^2 = 0.840$) factors, as well as the interaction between the two factors ($F = 130.927$, $p < 0.001$, $\eta_p^2 = 0.819$). For the NMF condition factor that reached a significance, SE values were significantly larger (mean difference $= 43.892 \pm 3.551$ Hz/ms) after the NMF algorithm was applied. For clarity, results of all pair-wise comparisons across the 11 nSweeps conditions are also summarized in Appendix F.

For TA, statistical results of the two-way ANOVA at 100 iterations demonstrated a significant difference for the nSweeps condition ($F = 188.744$, $p < 0.001$, $\eta_p^2 = 0.867$) and NMF condition ($F = 191.989$, $p < 0.001$, $\eta_p^2 = 0.869$) factors, as well as the interaction between the two factors ($F = 46.762$, $p < 0.001$, $\eta_p^2 = 0.617$). For the NMF condition factor that reached a significance, TA values were significantly larger (mean difference = 0.131 ± 0.009) after the NMF algorithm was applied. Finally, the results of all pair-wise comparisons across the 11 nSweeps conditions are also summarized in Appendix F.

Comparison of the NMF Performance Between Adults and Newborns (Aim 3b)

As a follow up to Aim 3a, the purpose of Aim 3b was to compare the adult and newborn NMF results in an attempt to better understand the differences, if any, between the two age groups (adults and neonates).

Figure 15 depicts the results of the NMF condition for adults and newborns. Beginning with the first panel, Panel A displays the FE results. For the adult NMF recordings, FE values were 7.583 ± 0.627 Hz at 1 iteration for 100 sweeps. With an increasing number of sweeps, FE values decreased and plateaued at about 3000 sweeps. For the newborn NMF recordings, similarities were present in the overall response trends as those observed in the adult NMF recordings, with two exceptions. One difference was that the FE values for newborns were higher at 100 sweeps (9.202 ± 0.827 Hz) than those observed in the adult NMF recordings. Another difference was that, approximately 5000-8000 sweeps were required to reach a plateau for newborns. As expected, a higher number of sweeps was required to reach a plateau in the newborns. When viewing the data from 5, 10, and 100 iteration results (Panels B, C, & D) remained consistent with the 1 iteration findings.

Panel E displays the results for the SE feature. When viewing the adult NMF recordings, SE values were largest at 100 sweeps (-96.157 ± 15.334 Hz/ms) for 1 iteration. With an increasing number of sweeps, SE values increased and became closer to zero Hz/ms. In the adult NMF, approximately 2000-4000 sweeps were needed to observe a plateau. When viewing the newborn NMF data, SE values were -142.206 ± 15.207 Hz/ms at 100 sweeps. For the newborn NMF, roughly 4000-8000 sweeps were needed to reach a plateau comparable to the adult NMF. As expected, a higher number of sweeps was required to reach a plateau in the newborns. Furthermore, the data from 5, 10, and 100 iterations (Panels F, G, & H) were similar to the 1 iteration results.

Finally, the results of the TA feature can be viewed in Panel I. For the adult NMF results, at 1 iteration and 100 sweeps TA values were 0.712 ± 0.042. With an increasing number of sweeps, TA values increased. For the adult NMF data, in order to observe a plateau in TA values roughly 2000-4000 sweeps were needed. For the newborn NMF recordings, similarities were present in the overall response trends as those observed in the adult NMF recordings, with two exceptions. One difference was that the FE values for newborns were higher at 100 sweeps (0.492 ± 0.052). Another difference was that, roughly 4000-6000 sweeps were required to reach a plateau in newborns. When comparing the adult NMF and newborn NMF results, it was apparent that the adult NMF required fewer sweeps in order to reach its plateau whereas the newborn results required more. When comparing the data from 5, 10, and 100 iterations (Panels J, K, & L), results remained consistent with 1 iteration.

Figure 15

Graphical presentations of the raw data extracts for the adult and newborn NMF conditions at 1, 5, 10 and 100 iterations for Frequency Error (first row), Slope Error (second row), and Tracking Accuracy (third row).

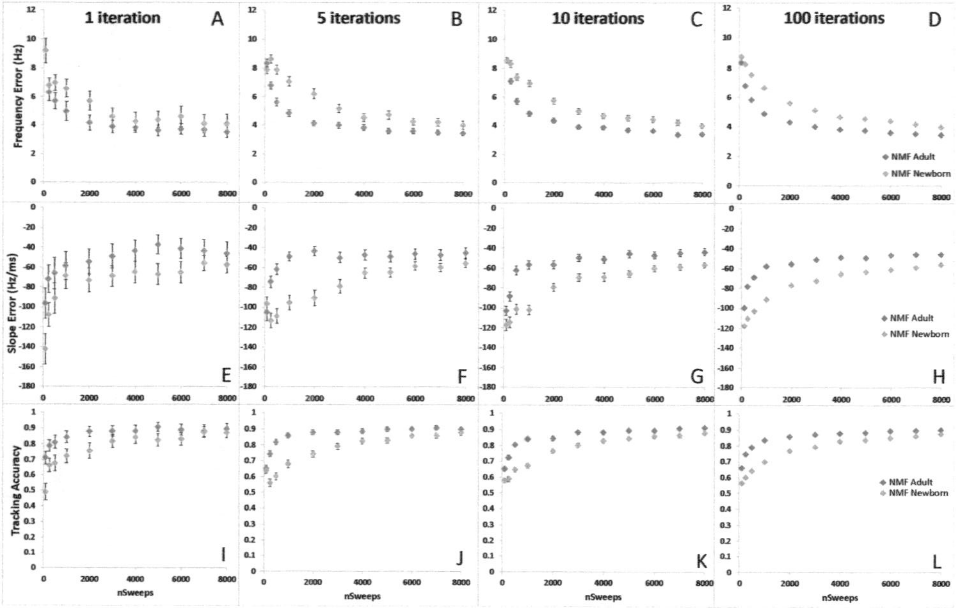

Linear Regression

Furthermore, a linear regression analysis was conducted between the adult and newborn NMF groups. These results can be viewed in Figure 16. Panel A shows the adult and newborn data for FE, Panel E depicts the adult and newborn data for SE, and Panel I display the adult and newborn data for TA. Equations for the linear regression are listed as follows

1 Iteration.

Adult Linear Regression for FE: $y = -0.0003x + 5.4143$ (127)

Newborn Linear Regression for FE $y = -0.0005x + 7.2077$ (128)

Adult Linear Regression for SE: $y = 0.0047x - 70.9$ (129)

Newborn Linear Regression for SE: $y = 0.0067x - 100.84$ (130)

Adult Linear Regression for TA: $y = 2E\text{-}05x + 0.7989$ (131)

Newborn Linear Regression for TA: $y = 3E\text{-}05x + 0.6463$ (132)

For FE, slope estimates for Equations 127 and 128 displayed differing values. The newborn NMF slope was steeper than the adult NMF group. This means that for the newborn group, FE values decreased at a faster rate compared to those observed in the adult group. In other words, the decrement for the newborn FE feature took a smaller number of sweeps (i.e., less time) to obtain a similar performance. When viewing the differing FE intercepts, values for the newborn group indicated a higher noise floor for the FE trend than that of the adult group.

When looking at the adult and newborn NMF SE Equations (129 & 130), there were differences between the slope estimates. The newborn NMF slope was steeper than the adult slope. Meaning that, the time it took for the newborn NMF to reach comparable performance to the adult NMF was shorter. Therefore, the SE trend of the newborn NMF recordings took a smaller number of sweeps to obtain a similar performance to the adult SE. When viewing the differing adult and newborn SE intercepts, values for the newborn group indicated a higher noise floor for the SE trend than those observed in the adult group.

For TA, the adult (Equation 131) and newborn (Equation 132) NMF slope estimates displayed differing values. The newborn NMF slope was steeper than the adult NMF slope. This means that for the newborn TA values increased at a faster rate compared to the adult TA values. In other words, the increment for the newborn TA took a smaller number of sweeps to obtain a similar performance when compared to the adult data. When viewing the differing intercepts, values for the adult group indicated a lower noise floor for the TA trend when compared to those observed in the newborn group.

5 Iterations.

$$\text{Adult Linear Regression for FE: } y = -0.0004x + 6.1808 \tag{133}$$

$$\text{Newborn Linear Regression for FE } y = -0.0006x + 7.7629 \tag{134}$$

$$\text{Adult Linear Regression for SE: } y = 0.004x - 69.657 \tag{135}$$

$$\text{Newborn Linear Regression for SE: } y = 0.007x - 104.15 \tag{136}$$

$$\text{Adult Linear Regression for TA: } y = 2\text{E-}05x + 0.7779 \tag{137}$$

$$\text{Newborn Linear Regression for TA: } y = 4\text{E-}05x + 0.6276 \tag{138}$$

When viewing the 5 iteration equations, the values and trends remained similar to the 1 iteration results.

10 Iterations.

$$\text{Adult Linear Regression for FE: } y = -0.0005x + 7.6348 \tag{139}$$

Newborn Linear Regression for FE y = -0.0004x + 5.6957 (140)

Adult Linear Regression for SE: y = 0.0075x - 106.53 (141)

Newborn Linear Regression for SE: y = 0.0052x - 76.421 (142)

Adult Linear Regression for TA: y = 2E-05x + 0.7608 (143)

Newborn Linear Regression for TA: y = 4E-05x + 0.6325 (144)

When viewing the 10 iteration equations, the values and trends remained similar to the 1 and 5 iteration results.

100 Iterations.

Adult Linear Regression for FE: y = -0.0005x + 6.2556 (145)

Newborn Linear Regression for FE y = -0.0006x + 7.6245 (146)

Adult Linear Regression for SE: y = 0.0046x - 74.54 (147)

Newborn Linear Regression for SE: y = 0.0071x - 103.46 (148)

Adult Linear Regression for TA: y = 2E-05x + 0.766 (149)

Newborn Linear Regression for TA: y = 4E-05x + 0.6358 (150)

When viewing the 100 iteration equations, the values and trends remained similar to the 1, 5, and 10 iteration results.

Figure 16

Graphical presentations of the linear regressions performed on the adult and newborn NMF conditions at 1, 5, 10 and 100 iterations for Frequency Error (first row), Slope Error (second row), and Tracking Accuracy (third row).

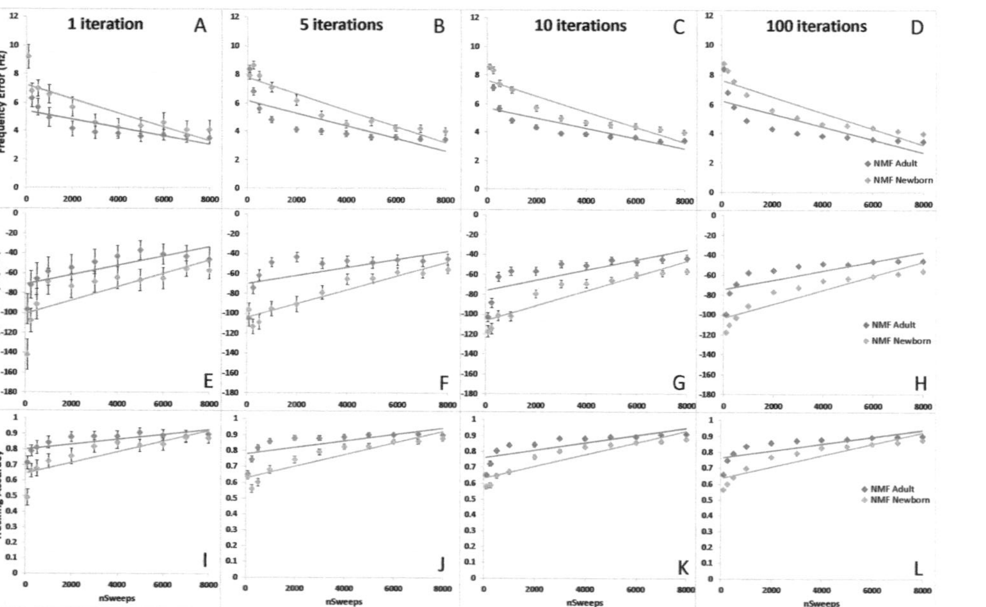

Exponential Modeling

Exponential modeling was conducted on the adult and newborn NMF groups. Results for the adult and newborn groups can be seen in Figure 17. To reiterate, Panel A depicts the FE results, Panel E depicts the SE results, and Panel I depicts the TA results. Each column displayed the varying iterations. Derived equations are listed as follows. A tabular depiction of the Tau, A_{noise}, and A_{as} values can be viewed in Appendix G.

1 Iteration.

Adult Exponential Modeling for FE: $A(n) = 6.693 \left(e^{-\frac{n}{1071.009}} \right) - 3.644$ (151)

Newborn Exponential Modeling for FE: $A(n) = 7.783 \left(e^{-\frac{n}{2050.346}} \right) - 4.104$ (152)

Adult Exponential Modeling for SE: $A(n) = -43.410 \left(1 - e^{-\frac{n}{1107.723}} \right) + 77.357$ (153)

Newborn Exponential Modeling for SE: $A(n) = -65.435 \left(1 - e^{-\frac{n}{282.656}} \right) + 142.206$ (154)

Adult Exponential Modeling for TA: $A(n) = 0.883 \left(1 - e^{-\frac{n}{457.287}} \right) - 0.712$ (155)

Newborn Exponential Modeling for TA: $A(n) = 0.873 \left(1 - e^{-\frac{n}{2047.339}} \right) - 0.634$ (156)

For FE, the exponential modeling Equations 151 and 152 displayed differing values. The noise amplitude (A_{noise}) values estimated for the adult and newborn groups were 6.693 Hz and 7.783 Hz, respectively. These values indicated that the A_{noise} estimates in the newborn group was 14% ([7.783 -6.693] / 7.783 = 14%) lower than that of the adult group. The τ values of the fitted curves for FE trends obtained in the adult and newborn groups were 1071 and 2050 sweeps, respectively. A noteworthy observation showed that the adult participants produced a smaller τ value than neonates. This finding indicated that the adult participants exhibited a faster improvement in FE values with increasing nSweeps when compared to the neonates. The asymptotic amplitude (A_{as}) for the adult and newborn groups were 3.644 Hz and 4.104 Hz, respectively. Similar FE plateau values indicated that the age group had little to no influence on the FE plateau.

For SE, the exponential modeling Equations 153 and 154 displayed a difference in values. The asymptotic amplitude (A_{AS}) values estimated for the adult and newborn groups were -43.410 Hz/ms and -65.435 Hz/ms, respectively. The differing SE plateau values indicated that the adult age group had little influence on the SE plateau. The τ values of the fitted curves for SE trends obtained in the adult and newborn groups were 1107 and 282 sweeps, respectively. An interesting observation showed that the newborn participants produced a smaller τ value than adults. This finding indicated that the newborn participants displayed a faster improvement SE with increasing nSweeps when compared to the adults. The noise amplitude (A_{noise}) values estimated for the adult and newborn groups were -77.357 Hz/ms and -142.206 Hz/ms, respectively. These values indicated that the A_{noise} estimates in the newborn group were 45% ([-142.206 + 77.357] / -142.206 = 45%) lower than that of the adult group.

For TA, the exponential modeling Equations 155 and 156 displayed varying values. The asymptotic amplitude (A_{AS}) values estimated for the adult and newborn groups were 0.883 and 0.873, respectively. Similar TA plateau values indicated that the age group had little to no influence on the TA plateau values. The τ values of the fitted curves for TA trends obtained in the adult and newborn groups were 457 and 2047 sweeps, respectively. The adult participants produced a smaller τ value than the newborns. This finding indicated that the adult participants had a faster improvement in tracking accuracy with an increasing number of sweeps when compared to those obtained from the newborns. The noise amplitude (A_{noise}) values estimated for the adult and newborn groups were 0.712 and 0.634, respectively. These values indicated that the A_{noise} estimates in the adult group was 11% ([0.712 – 0.634] / 0.712 = 11%) lower than that of the newborn group.

5 Iterations.

Adult Exponential Modeling for FE: $A(n) = 8.356\left(e^{-\frac{n}{463.151}}\right) - 3.585$ (157)

Newborn Exponential Modeling for FE: $A(n) = 8.681\left(e^{-\frac{n}{2241.782}}\right) - 4.030$ (158)

Adult Exponential Modeling for SE: $A(n) = -47.048\left(1 - e^{-\frac{n}{256.498}}\right) + 104.963$ (159)

Newborn Exponential Modeling for SE: $A(n) = -55.460\left(1 - e^{-\frac{n}{3446.356}}\right) + 116.323$ (160)

Adult Exponential Modeling for TA: $A(n) = 0.896\left(1 - e^{-\frac{n}{350.057}}\right) - 0.651$ (161)

Newborn Exponential Modeling for TA: $A(n) = 0.871\left(1 - e^{-\frac{n}{2215.412}}\right) - 0.557$ (162)

When viewing the 5 iteration equations, the A_{Noise}, τ, and A_{AS} values for FE and TA remained similar to those observed in the 1 iteration results. For SE, its A_{Noise} and A_{AS} values at 5 iterations were also consistent with the 1 iteration results. However, for the SE τ values, the exponential modeling Equations 159 and 160 displayed differing τ values, when compared to 1 iteration. Specifically, the SE τ values of the fitted curves for the adult and newborn participants were 256 and 3446 sweeps, respectively. It was important to note that the NMF condition produced a larger τ value than the ORI condition. This finding indicated that the ORI condition showed a faster improvement in SE with an increasing number of sweeps compared to the NMF condition.

10 Iterations.

Adult Exponential Modeling for FE: $A(n) = 8.067\left(e^{-\frac{n}{648.570}}\right) - 3.670$ (163)

Newborn Exponential Modeling for FE: $A(n) = 8.571\left(e^{-\frac{n}{1839.548}}\right) - 4.177$ (164)

Adult Exponential Modeling for SE: $A(n) = -47.028\left(1 - e^{-\frac{n}{499.799}}\right) + 103.203$ (165)

Newborn Exponential Modeling for SE: $A(n) = -57.654\left(1 - e^{-\frac{n}{2315.143}}\right) + 117.387$ (166)

Adult Exponential Modeling for TA: $A(n) = 0.890\left(1 - e^{-\frac{n}{417.735}}\right) - 0.652$ (167)

Newborn Exponential Modeling for TA: $A(n) = 0.867\left(1 - e^{-\frac{n}{2046.337}}\right) - 0.578$ (168)

When viewing the 10 iteration equations, the values remained similar to the 5 iteration results.

100 Iterations.

Adult Exponential Modeling for FE: $A(n) = 7.743\left(e^{-\frac{n}{700.376}}\right) - 3.721$ (169)

Newborn Exponential Modeling for FE: $A(n) = 8.840\left(e^{-\frac{n}{1598.838}}\right) - 4.280$ (170)

Adult Exponential Modeling for SE: $A(n) = -48.847\left(1 - e^{-\frac{n}{440.170}}\right) + 99.873$ (171)

Newborn Exponential Modeling for SE: $A(n) = -60.150\left(1 - e^{-\frac{n}{1636.003}}\right) + 115.059$ (172)

Adult Exponential Modeling for TA: $A(n) = 0.881\left(1 - e^{-\frac{n}{444.510}}\right) - 0.661$ (173)

Newborn Exponential Modeling for TA: $A(n) = 0.854\left(1 - e^{-\frac{n}{1662.349}}\right) - 0.576$ (174)

When viewing the 100 iteration equations, the values remained similar to the 5 and 10 iteration results.

Figure 17

Graphical displays of the exponential modeling performed on the NMF data for the adult and newborn groups at 1, 5, 10 and 100 iterations for Frequency Error (first row), Slope Error (second row), and Tracking Accuracy (third row).

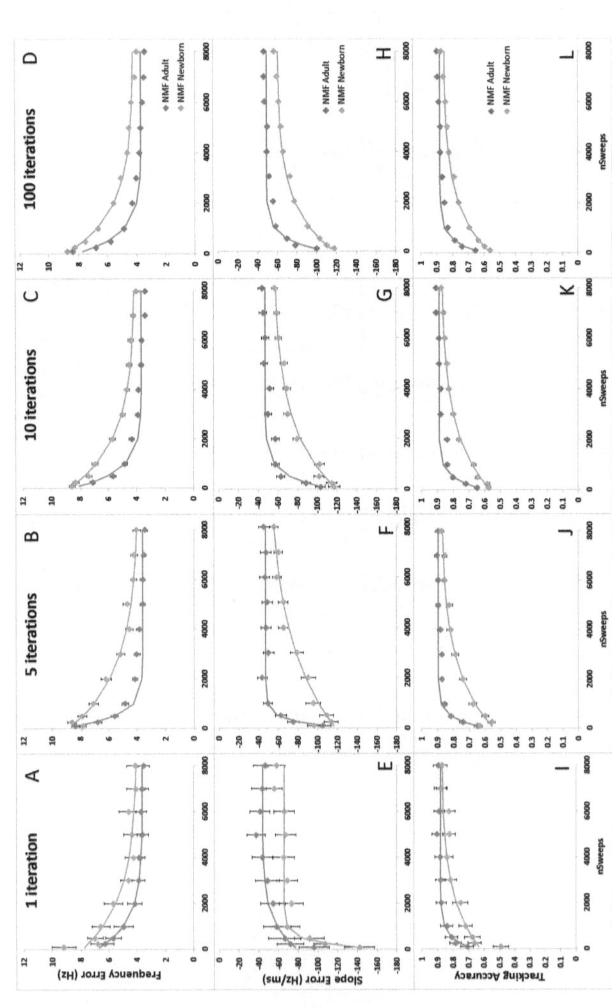

Three-way ANOVA

1 Iteration.

For FE, statistical results of a three-way ANOVA at 1 iteration demonstrated a significant difference for the nSweeps condition ($F = 78.831$, $p < 0.001$, $\eta_p^2 = 0.576$) and NMF condition ($F = 178.137$, $p < 0.001$, $\eta_p^2 = 0.754$) factors, as well as the effect of age ($F = 4.048$, $p = 0.049$, $\eta_p^2 = 0.418$). A significant difference was also observed for the interactions between the nSweeps condition and the NMF condition factors ($F = 34.711$, $p < 0.001$, $\eta_p^2 = 0.374$). This interaction means that, for FE, the NMF algorithm affects the number of sweeps needed to observe a response. A significant difference was also observed between the NMF condition and age groups ($F = 9.227$, $p < 0.004$, $\eta_p^2 = 0.137$). Simply put, the NMF algorithm affects the FE values for both adults and newborns, where adult FE values were smaller when the NMF algorithm was applied than those obtained in newborns. There was no significant difference observed between the nSweeps condition and the age groups ($F = 1.540$, $p > 0.001$, $\eta_p^2 = 0.026$), as well as the interaction between all three factors ($F = 1.698$, $p = 0.134$, $\eta_p^2 = 0.028$). For the NMF condition factor that reached a significance, FE values were significantly smaller (mean difference = -2.058 ± 0.154Hz) for the adult participants than those observed in for the newborn participants. For the age factor, FE values for adults were significantly smaller (mean difference = -1.413 ± 0.702 Hz) than those obtained in the newborn group. For clarity, results of all pair-wise comparisons across the 11 nSweeps conditions are summarized in Appendix C.

For SE, statistical results of a three-way ANOVA at 1 iteration demonstrated a significant difference for the nSweeps condition ($F = 35.637$, $p < 0.001$, $\eta_p^2 = 0.381$) and

NMF condition ($F = 120.001$, $p < 0.001$, $\eta_p^2 = 0.674$) factors, as well as the interaction between the age groups ($F = 4.657$, $p = 0.035$, $\eta_p^2 = 0.074$). A significant difference was also observed for the interactions between the nSweeps condition and the NMF condition factors ($F = 23.736$, $p < 0.001$, $\eta_p^2 = 0.290$). This interaction means that, for SE, the NMF algorithm affects the number of sweeps needed to observe a response. There was no significant difference observed between the nSweeps condition and the age groups ($F = 0.809$, $p > 0.001$, $\eta_p^2 = 0.014$), the NMF condition and age groups ($F = 2.327$, $p = 0.133$, $\eta_p^2 = 0.039$), as well as the interaction between all three factors ($F = 0.578$, $p = 0.718$, $\eta_p^2 = 0.010$). For the NMF factor that reached a significance, SE values were significantly larger (mean difference = 38.197 ± 3.487 Hz/ms) for the adult participants than those observed in for the newborn participants. For the age factor, SE values for adults were significantly larger (mean difference = 28.521 ± 13.216 Hz/ms) than those obtained in the newborn group. Furthermore, the results of all pair-wise comparisons for the 11 nSweeps conditions are also summarized in Appendix C.

For TA, statistical results of the three-way ANOVA at 1 iteration demonstrated a significant difference for the nSweeps condition ($F = 52.193$, $p < 0.001$, $\eta_p^2 = 0.474$) and NMF condition ($F = 100.592$, $p < 0.001$, $\eta_p^2 = 0.634$) factors, as well as the interaction between the age groups ($F = 6.059$, $p = 0.017$, $\eta_p^2 = 0.095$). A significant difference was also observed for the interactions between the nSweeps condition and the NMF condition factors ($F = 23.736$, $p < 0.001$, $\eta_p^2 = 0.290$). This interaction means that, for TA, the NMF algorithm affects the number of sweeps needed to observe a response. There was no significant difference observed between the nSweeps condition and the age groups ($F = $

1.940, $p = 0.075$, $\eta_p^2 = 0.032$) and the NMF condition and age groups ($F = 0.893$, $p = 0.349$, $\eta_p^2 = 0.015$), as well as the interaction between all three factors ($F = 2.713$, $p = 0.015$, $\eta_p^2 = 0.051$). For the NMF condition factor that reached a significance, FE values were significantly smaller (mean difference $= 0.108 \pm 0.011$) for the adult participants than those observed in for the newborn participants. For the age factor, FE values for adults were significantly smaller (mean difference $= 0.101 \pm 0.041$) than those obtained in the newborn group. Finally, the results of all pair-wise comparisons for the 11 nSweeps conditions are also summarized in Appendix C.

5 Iterations.

For FE, statistical results of the three-way ANOVA at 5 iterations demonstrated a significant difference for the nSweeps condition ($F = 194.446$, $p < 0.001$, $\eta_p^2 = 0.770$) and NMF condition ($F = 280.436$, $p < 0.001$, $\eta_p^2 = 0.829$) factors, as well as the interaction between the age groups ($F = 5.335$, $p = 0.24$, $\eta_p^2 = 0.084$). A significant difference was also observed for the interactions between the nSweeps condition and the NMF condition factors ($F = 93.931$, $p < 0.001$, $\eta_p^2 = 0.618$). This interaction means that, for FE, the NMF algorithm affects the number of sweeps needed to observe a response. There was also a significant difference between the nSweeps condition and the age groups ($F = 8.111$, $p < 0.001$, $\eta_p^2 = 0.123$). Thus, as the number of sweeps increases for adults FE values were smaller than the newborn values. Finally, a difference was also observed for the NMF condition and age groups ($F = 8.538$, $p = 0.005$, $\eta_p^2 = 0.128$). Simply put, the NMF algorithm affects the FE values for both adults and newborns, where adult FE values were smaller when the NMF algorithm was applied than those obtained in newborns.

There was no significant difference observed for interaction between all three factors (F = 2.175, p = 0.067, η_p^2 = 0.036). For the NMF condition factor that reached a significance, FE values were significantly smaller (mean difference = -2.143 ± 0.128 Hz) for the adult participants than those observed in for the newborn participants. For the age factor, FE values for adults were significantly smaller (mean difference = -1.545 ± 0.669 Hz) than those obtained in the newborn group. Finally, the results of all pair-wise comparisons across the 11 nSweeps conditions are summarized in Appendix D.

For SE, statistical results of the three-way ANOVA at 5 iterations demonstrated a significant difference for the nSweeps condition (F = 66.398, p < 0.001, η_p^2 = 0.534) and NMF condition (F = 164.990, p < 0.001, η_p^2 = 0.740) factors, as well as the interaction between the age groups (F = 6.013, p = 0.017, η_p^2 = 0.094). A significant difference was also observed for the interactions between the nSweeps condition and the NMF condition factors (F = 95.497, p < 0.001, η_p^2 = 0.622). This interaction means that, for SE, the NMF algorithm affects the number of sweeps needed to observe a response. A significant difference was also observed for the nSweeps condition and the age groups (F = 6.227, p < 0.001, η_p^2 = 0.097). Therefore, as the number of sweeps increases for adults SE values were larger than the newborn values. Finally, a significant difference was observed between the NMF condition and age groups (F = 5.627, p = 0.021, η_p^2 = 0.089). Simply put, the NMF algorithm affects the SE values for both adults and newborns, where adult SE values were larger when the NMF algorithm was applied than those obtained in newborns. There was no significant difference observed for interaction between all three factors (F = 1.810, p = 0.117, η_p^2 = 0.030). For the NMF condition factor that reached a significance, SE values were significantly larger (mean difference = 38.060 ± 2.963

Hz/ms) for the adult participants than those observed in for the newborn participants. For

the age factor, SE values for adults were significantly larger (mean difference = 31.499 ±

12.846 Hz/ms) than those obtained in the newborn group. Furthermore, the results of all

pair-wise comparisons across the 11 nSweeps conditions are also summarized in

Appendix D.

For TA, statistical results of the three-way ANOVA at 5 iterations demonstrated a

significant difference for the nSweeps condition ($F = 136.153$, $p < 0.001$, $\eta_p^2 = 0.701$) and

NMF condition ($F = 183.896$, $p < 0.001$, $\eta_p^2 = 0.760$) factors, as well as the interaction

between the age groups ($F = 8.252$, $p = 0.006$, $\eta_p^2 = 0.125$). A significant difference was

also observed for the interactions between the nSweeps condition and the NMF condition

factors ($F = 35.310$, $p < 0.001$, $\eta_p^2 = 0.378$). This interaction means that, for TA, the NMF

algorithm affects the number of sweeps needed to observe a response. The nSweeps

condition and the age groups also showed a significant difference ($F = 7.015$, $p < 0.001$,

$\eta_p^2 = 0.108$). Therefore, as the number of sweeps increases for adult's TA values were

larger than the newborn values. There was also a significance observed in the interaction

between all three factors ($F = 2.416$, $p = 0.037$, $\eta_p^2 = 0.040$). No significant difference

was observed in the interaction between the NMF condition and age groups ($F = 3.321$, p

$= 0.074$, $\eta_p^2 = 0.054$). For the NMF condition factor that reached a significance, TA

values were significantly larger (mean difference = 0.115 ± 0.009) for the adult

participants than those observed in for the newborn participants. For the age factor, TA

values for adults were significantly larger (mean difference = 0.112 ± 0.039) than those obtained in the newborn group. For clarity, results of all pair-wise comparisons across the 11 nSweeps conditions are also summarized in Appendix D.

10 Iterations.

For FE, statistical results of the three-way ANOVA at 10 iterations demonstrated a significant difference for the nSweeps condition ($F = 256.585$, $p < 0.001$, $\eta_p^2 = 0.816$) and NMF condition ($F = 327.895$, $p < 0.001$, $\eta_p^2 = 0.850$) factors, as well as the interaction between the age groups ($F = 4.486$, $p = 0.038$, $\eta_p^2 = 0.072$). A significant difference was also observed for the interactions between the nSweeps condition and the NMF condition factors ($F = 120.934$, $p < 0.001$, $\eta_p^2 = 0.676$). This interaction means that, for FE, the NMF algorithm affects the number of sweeps needed to observe a response. There was also a significant difference between the nSweeps condition and the age groups $F = 4.777$, $p < 0.001$, $\eta_p^2 = 0.816$). Thus, as the number of sweeps increases for adults FE values were smaller than the newborn values. Finally, a difference was also observed for the NMF condition and age groups ($F = 9.501$, $p = 0.003$, $\eta_p^2 = 0.141$). Simply put, the NMF algorithm affects the FE values for both adults and newborns, where adult FE values were smaller when the NMF algorithm was applied than those obtained in newborns. There was no significant difference observed for interaction between all three factors ($F = 1.969$, $p = 0.120$, $\eta_p^2 = 0.033$). For the NMF condition factor that reached a significance, FE values were significantly smaller (mean difference = -2.113 $\pm$ 0.117 Hz) for the adult participants than those observed in for the newborn participants. For the age factor, FE values for adults were significantly smaller (mean difference = -1.420 $\pm$ 0.670 Hz) than those obtained in the newborn group. For clarity, results of all pair-wise comparisons across the 11 nSweeps conditions are summarized in Appendix E.

For SE, statistical results of the three-way ANOVA at 10 iterations demonstrated a significant difference for the nSweeps condition ($F = 109.903$, $p < 0.001$, $\eta_p^2 = 0.655$) and NMF condition ($F = 158.557$, $p < 0.001$, $\eta_p^2 = 0.732$) factors, as well as the interaction between the age groups ($F = 5.893$, $p = 0.018$, $\eta_p^2 = 0.092$). A significant difference was also observed for the interactions between the nSweeps condition and the NMF condition factors ($F = 111.569$, $p < 0.001$, $\eta_p^2 = 0.658$). This interaction means that, for SE, the NMF algorithm affects the number of sweeps needed to observe a response. A significant difference was also observed for the nSweeps condition and the age groups ($F = 4.533$, $p < 0.001$, $\eta_p^2 = 0.076$). Therefore, as the number of sweeps increases for adults SE values were larger than the newborn values. Finally, a significant difference was observed between the NMF condition and age groups ($F = 6.868$, $p = 0.012$, $\eta_p^2 = 0.103$). Simply put, the NMF algorithm affects the SE values for both adults and newborns, where adult SE values were larger when the NMF algorithm was applied than those obtained in newborns. There was no significant difference observed for interaction between all three factors ($F = 2.515$, $p = 0.055$, $\eta_p^2 = 0.042$). For the NMF condition factor that reached a significance, SE values were significantly larger (mean difference = 36.752 ± 2.919 Hz/ms) for the adult participants than those observed in for the newborn participants. For the age factor, SE values for adults were significantly larger (mean difference = 29.805 ± 12.278 Hz/ms) than those obtained in the newborn group. Finally, the results of all pair-wise comparisons across the 11 nSweeps conditions are also summarized in Appendix E.

For TA, statistical results of the three-way ANOVA at 10 iterations demonstrated a significant difference for the nSweeps condition ($F = 180.892, p < 0.001, \eta_p^2 = 0.757$) and NMF condition ($F = 233.955, p < 0.001, \eta_p^2 = 0.801$) factors, as well as the interaction between the age groups ($F = 6.403, p = 0.014, \eta_p^2 = 0.099$). A significant difference was also observed for the interactions between the nSweeps condition and the NMF condition factors ($F = 55.561, p < 0.001, \eta_p^2 = 0.489$). This interaction means that, for TA, the NMF algorithm affects the number of sweeps needed to observe a response. The nSweeps condition and the age groups also showed a significant difference ($F = 6.513, p < 0.001, \eta_p^2 = 0.101$). Therefore, as the number of sweeps increases for adult's TA values were larger than the newborn values. Finally, a significant difference was observed between the NMF condition and age groups ($F = 4.633, p = 0.036, \eta_p^2 = 0.074$). Simply put, the NMF algorithm affects the TA values for both adults and newborns, where adult TA values were larger when the NMF algorithm was applied than those obtained in newborns. There was also a significance observed in the interaction between all three factors ($F = 3.211, p = 0.017, \eta_p^2 = 0.052$). For the NMF condition factor that reached a significance, TA values were significantly larger (mean difference $= 0.116 \pm 0.008$) for the adult participants than those observed in for the newborn participants. For the age factor, TA values for adults were significantly larger (mean difference $= 0.099 \pm 0.039$) than those obtained in the newborn group. Furthermore, the results of all pair-wise comparisons across the 11 nSweeps conditions are also summarized in Appendix E.

100 Iterations.

For FE, statistical results of the three-way ANOVA at 100 iterations demonstrated a significant difference for the nSweeps condition ($F = 360.072$, $p < 0.001$, $\eta_p^2 = 0.861$) and NMF condition ($F = 331.516$, $p < 0.001$, $\eta_p^2 = 0.851$) factors, as well as the interaction between the age groups ($F = 4.329$, $p = 0.042$, $\eta_p^2 = 0.069$).. A significant difference was also observed for the interactions between the nSweeps condition and the NMF condition factors ($F = 186.341$, $p < 0.001$, $\eta_p^2 = 0.763$). This interaction means that, for FE, the NMF algorithm affects the number of sweeps needed to observe a response. There was also a significant difference between the nSweeps condition and the age groups ($F = 7.621$, $p < 0.001$, $\eta_p^2 = 0.116$). Thus, as the number of sweeps increases for adults FE values were smaller than the newborn values. Finally, a difference was also observed for the NMF condition and age groups ($F = 10.698$, $p = 0.002$, $\eta_p^2 = 0.156$). Simply put, the NMF algorithm affects the FE values for both adults and newborns, where adult FE values were smaller when the NMF algorithm was applied than those obtained in newborns. There was also a significance observed in the interaction between all three factors ($F = 6.219$, $p = 0.002$, $\eta_p^2 = 0.097$). For the NMF condition factor that reached a significance, FE values were significantly smaller (mean difference = -2.098 ± 0.115 Hz) for the adult participants than those observed in for the newborn participants. For the age factor, FE values for adults were significantly smaller (mean difference = -1.403 ± 0.674 Hz) than those obtained in the newborn group. Lastly, the results of all pair-wise comparisons across the 11 nSweeps conditions are summarized in Appendix F.

For SE, statistical results of the three-way ANOVA at 100 iterations demonstrated a significant difference for the nSweeps condition ($F = 206.411$, $p < 0.001$, $\eta_p^2 = 0.781$) and NMF condition ($F = 216.464$, $p < 0.001$, $\eta_p^2 = 0.789$) factors, as well as the interaction between the age groups ($F = 5.379$, $p = 0.024$, $\eta_p^2 = 0.085$). A significant difference was also observed for the interactions between the nSweeps condition and the NMF condition factors ($F = 228.250$, $p < 0.001$, $\eta_p^2 = 0.797$). This interaction means that, for SE, the NMF algorithm affects the number of sweeps needed to observe a response. A significant difference was also observed for the nSweeps condition and the age groups ($F = 6.297$, $p = 0.004$, $\eta_p^2 = 0.098$). Therefore, as the number of sweeps increases for adults SE values were larger than the newborn values. Finally, a significant difference was observed between the NMF condition and age groups ($F = 8.737$, $p = 0.005$, $\eta_p^2 = 0.131$). Simply put, the NMF algorithm affects the SE values for both adults and newborns, where adult SE values were larger when the NMF algorithm was applied than those obtained in newborns. There was also a significance observed in the interaction between all three factors ($F = 4.566$, $p = 0.016$, $\eta_p^2 = 0.073$). For the NMF condition factor that reached a significance, SE values were significantly larger (mean difference = 36.549 ± 2.484 Hz/ms) for the adult participants than those observed in for the newborn participants. For the age factor, SE values for adults were significantly larger (mean difference = 28.064 ± 12.100 Hz/ms) than those obtained in the newborn group. Furthermore, the results of all pair-wise comparisons across the 11 nSweeps conditions are also summarized in Appendix F.

For TA, statistical results of the three-way ANOVA at 100 iterations demonstrated a significant difference for the nSweeps condition ($F = 274.226$, $p < 0.001$, $\eta_p^2 = 0.825$) and NMF condition ($F = 267.564$, $p < 0.001$, $\eta_p^2 = 0.822$) factors, as well as the interaction between the age groups ($F = 6.216$, $p = 0.016$, $\eta_p^2 = 0.097$). A significant difference was also observed for the interactions between the nSweeps condition and the NMF condition factors ($F = 98.576$, $p < 0.001$, $\eta_p^2 = 0.630$). This interaction means that, for TA, the NMF algorithm affects the number of sweeps needed to observe a response. The nSweeps condition and the age groups also showed a significant difference ($F = 9.327$, $p < 0.001$, $\eta_p^2 = 0.139$). Therefore, as the number of sweeps increases for adult's TA values were larger than the newborn values. Finally, a significant difference was observed between the NMF condition and age groups ($F = 6.337$, $p = 0.015$, $\eta_p^2 = 0.098$). Simply put, the NMF algorithm affects the TA values for both adults and newborns, where adult TA values were larger when the NMF algorithm was applied than those obtained in newborns. There was also a significance observed in the interaction between all three factors ($F = 9.069$, $p < 0.001$, $\eta_p^2 = 0.135$). For the NMF condition factor that reached a significance, TA values were significantly larger (mean difference = 0.114 ± 0.007) for the adult participants than those observed in for the newborn participants. For the age factor, TA values for adults were significantly larger (mean difference = 0.098 ± 0.039) than those obtained in the newborn group. Finally, the results of all pair-wise comparisons across the 11 nSweeps conditions are also summarized in Appendix F.

Evaluation of the NMF Performance for Newborns (Aim 4a)

Finally, the purpose of aim 4a was to evaluate the "apparent" and "hidden" components of an FFR response in neonates.

Figure 18 depicts the results of the hidden components extracted from the newborn recordings. Panel A displays the hFE results. For 1 iteration, the hidden component was smallest at 100 sweeps (-5.197 ± 0.785 Hz). With an increasing number of sweeps, hFE values began to increase. At roughly 6000 sweeps the hidden component began to taper and became very closer to zero Hz. Results of this investigation demonstrated that the NMF algorithm was able to uncover a substantial amount of the hFE components. This phenomenon was particularly apparent when 100-6000 sweeps were included. When viewing data from 5, 10, and 100 iterations (Panels B, C, & D), results were consistent with those observed at 1 iteration.

Panel E displays the hSE results. For 1 iteration, the hidden component was largest at 100 sweeps (122.904 ± 22.800 Hz/ms). With an increasing number of sweeps, hSE began to decrease and became close to zero Hz/ms. From about 6000 sweeps, the hSE component plateaued. Results from the investigation indicated that the NMF algorithm was able to reveal a substantial amount of hSE values at 100-6000 sweeps. These hSE values were previously unavailable when the NMF algorithm was not applied to the recordings. When viewing data from 5, 10, and 100 iterations (Panels F, G, & H), results were consistent with those observed at 1 iteration.

The results of the hTA feature can be viewed in Panel I. For 1 iteration, the hidden component was largest at 100 sweeps (0.167 ± 0.057). With an increasing number of sweeps, the hTA values began to decrease. From 6000 sweeps, the hidden component began to diminish and became very close to zero. In other words, a substantial performance was shown from 100-6000 sweeps in the results of the NMF algorithm for the hidden TA features. When comparing the data from 5, 10, and 100 iterations (J, K, &

L), results remained consistent with those at 1 iteration.

Figure 18

Graphical presentations display the raw data extracts for the newborn participants hidden component at 1, 5, 10 and 100 iterations for Hidden Frequency Error (first row), Hidden Slope Error (second row), and Hidden Tracking Accuracy (third row)

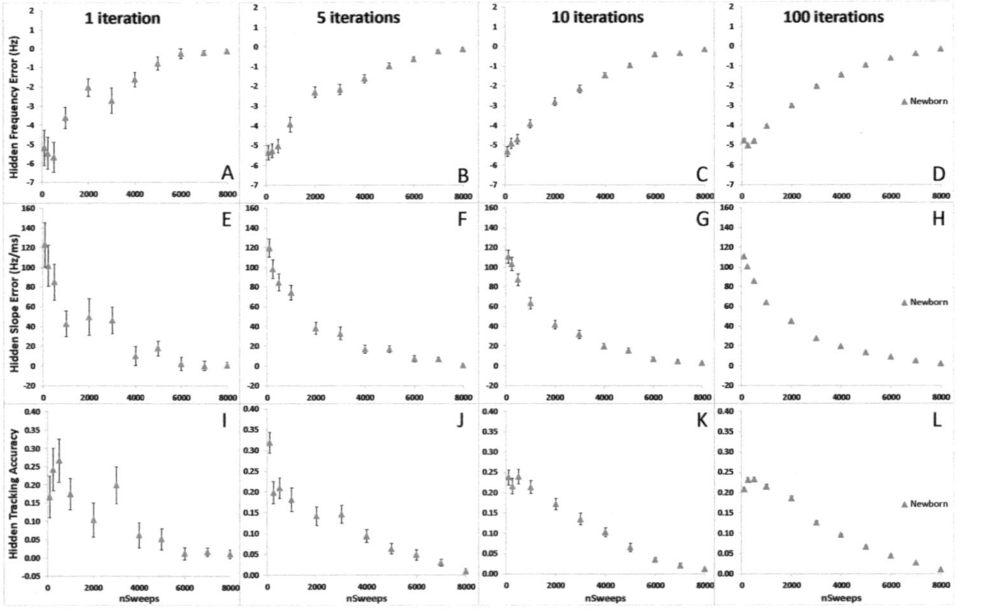

Linear Regression

Results of the linear regression analysis for the newborn hidden component can be seen in Figure 19. Similarly, to the previous linear regression analysis in this study Panel A depicts the hFE results, Panel E depicts the hSE results, and Panel I depicts the hTA results. Final equations can be observed below.

1 Iteration.

hFE Linear Regression: $y = 0.0007x + 4.9126$ (175)

hSE Linear Regression: $y = -0.0135x - 88.817$ (176)

hTA Linear Regression: $y = =3\text{E-05}x - 0.2182$ (177)

Slope estimates for Equation 175 depicts the hFE results. The hFE slope confirmed an incremental trend for hFE as nSweeps increased. The slope for hSE (shown in Equation 176) indicated a decremental trend for hSE as nSweeps increased. Finally, the hTAs (Equation 177) indicated a decreasing trend in hTA as nSweeps increased.

5 Iterations.

hFE Linear Regression: $y = 0.0007x - 4.8006$ (178)

hSE Linear Regression: $y = -0.0134x + 90.133$ (179)

hTA Linear Regression: $y = -3\text{E-05}x + 0.2331$ (180)

When viewing the 5 iteration equations, the values and trends remained similar to the 1 iteration results.

10 Iterations.

hFE Linear Regression: $y = 0.0007x - 4.6985$ (181)

hSE Linear Regression: $y = -0.0131x + 88.188$ (182)

hTA Linear Regression: $y = -3\text{E-05}x + 0.2356$ (183)

When viewing the 10 iteration equations, the values and trends remained similar to the 1 and 5 iteration results.

100 Iterations.

hFE Linear Regression: $y = 0.0006x - 4.6498$ (184)

hSE Linear Regression: $y = -0.0129x + 87.268$ (185)

hTA Linear Regression: $y = -3E\text{-}05x + 0.2311$ (186)

When viewing the 100 iteration equations, the values and trends remained similar to the 1, 5, and 10 iteration results.

Figure 19

Graphical depictions of the linear regressions for the newborn hidden component at 1, 5, 10 and 100 iterations for Hidden Frequency Error (first row), Hidden Slope Error (second row), and Hidden Tracking Accuracy (third row).

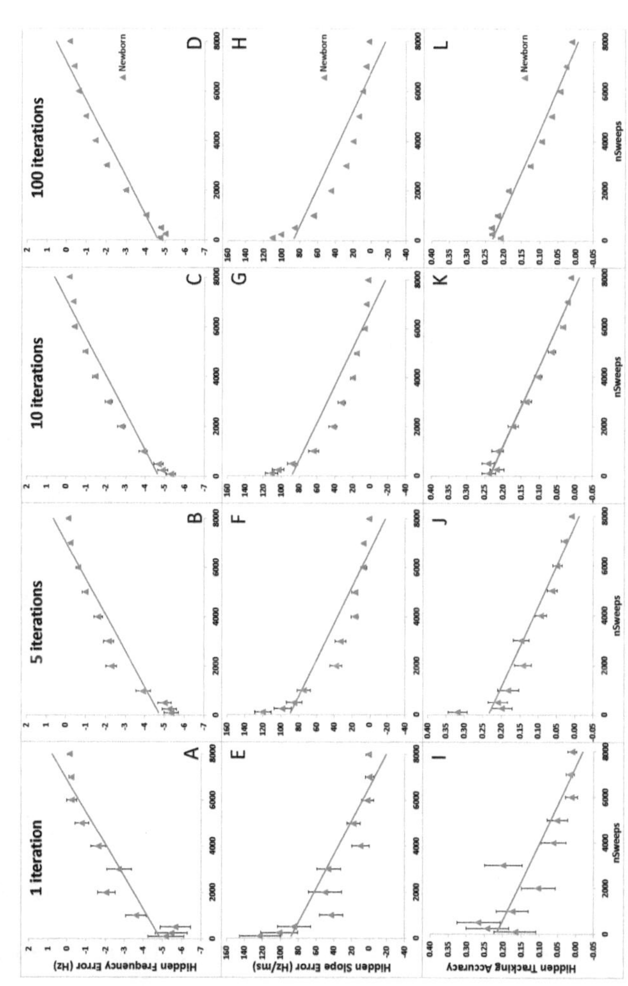

Exponential Modeling

Results of exponential modeling for the newborn hidden component can be seen in Figure 20. In line with the previous figures, Panel A depicts the FE results, Panel E depicts the SE results, and Panel I depicts the TA results. Each column in the figure displayed 1, 5, 10, and 100 iterations. Equations are listed by iteration as followed. A tabular depiction of the Tau, A_{noise}, and A_{as} values can be viewed in Appendix G.

1 Iterations.

$$\text{Newborn Exponential Modeling for hFE: } A(n) = -0.061\left(1 - e^{-\frac{n}{2968.763}}\right) + 5.655 \qquad (187)$$

$$\text{Newborn Exponential Modeling for hSE: } A(n) = 108.779\left(e^{-\frac{n}{2178.090}}\right) - 0.576 \qquad (188)$$

$$\text{Newborn Exponential Modeling for hTA: } A(n) = 0.010\left(e^{-\frac{n}{3136.437}}\right) - 0.240 \qquad (189)$$

The exponential modeling Equations 187, 188, and 189 indicated the overall trends for hFE, hSE, and hTA. The asymptotic amplitude (A_{AS}) for the hFE trend was -0.061 Hz, the A_{AS} value for the hSE trend was 0.576 Hz/ms, and for hTA the A_{AS} value was 0.240. The τ values for the hFE, hSE, and hTA trends were 2968, 2178, and 3136 sweeps, respectively. The noise amplitude (A_{noise}) value for the hFE trend was -5.655 Hz. For the hSE trend, the A_{noise} value was 108.779 Hz/ms. Furthermore, the noise amplitude for the hTA feature was 0.010. These values aided in the depiction of the hidden components' overall trends and plateaus within the exponential model.

5 Iterations.

$$\text{Newborn Exponential Modeling for hFE: } A(n) = -0.123\left(1 - e^{-\frac{n}{3783.525}}\right) + 5.376 \qquad (190)$$

$$\text{Newborn Exponential Modeling for hSE: } A(n) = 105.366\left(e^{-\frac{n}{2055.145}}\right) - 3.415 \qquad (191)$$

$$\text{Newborn Exponential Modeling for hTA: } A(n) = 0.013\left(e^{-\frac{n}{7645.131}}\right) - 0.225 \qquad (192)$$

When viewing the 5 iteration equations, the values remained similar to the 1 iteration results.

10 Iterations.

Newborn Exponential Modeling for hFE: $A(n) = -0.177\left(1 - e^{-\frac{n}{3460.903}}\right) + 5.310$ (193)

Newborn Exponential Modeling for hSE: $A(n) = 110.592\left(e^{-\frac{n}{1884.749}}\right) - 3.797$ (194)

Newborn Exponential Modeling for hTA: $A(n) = -0.001\left(e^{-\frac{n}{10667.208}}\right) - 0.248$ (195)

When viewing the 10 iteration equations, the values remained similar to the 1 and 5 iteration results.

100 Iterations.

Newborn Exponential Modeling for hFE: $A(n) = -0.161\left(1 - e^{-\frac{n}{3588.506}}\right) + 5.252$ (196)

Newborn Exponential Modeling for hSE: $A(n) = 107.736\left(e^{-\frac{n}{2078.514}}\right) - 4.515$ (197)

Newborn Exponential Modeling for hTA: $A(n) = 0.011\left(e^{-\frac{n}{6445.364}}\right) - 0.248$ (198)

When viewing the 100 iteration equations, the values remained similar to the 1, 5, and 10 iteration results.

Figure 20

Graphical displays of the exponential modeling performed on the newborn hidden component at 1, 5, 10 and 100 iterations for Hidden Frequency Error (first row), Hidden Slope Error (second row), and Hidden Tracking Accuracy (third row).

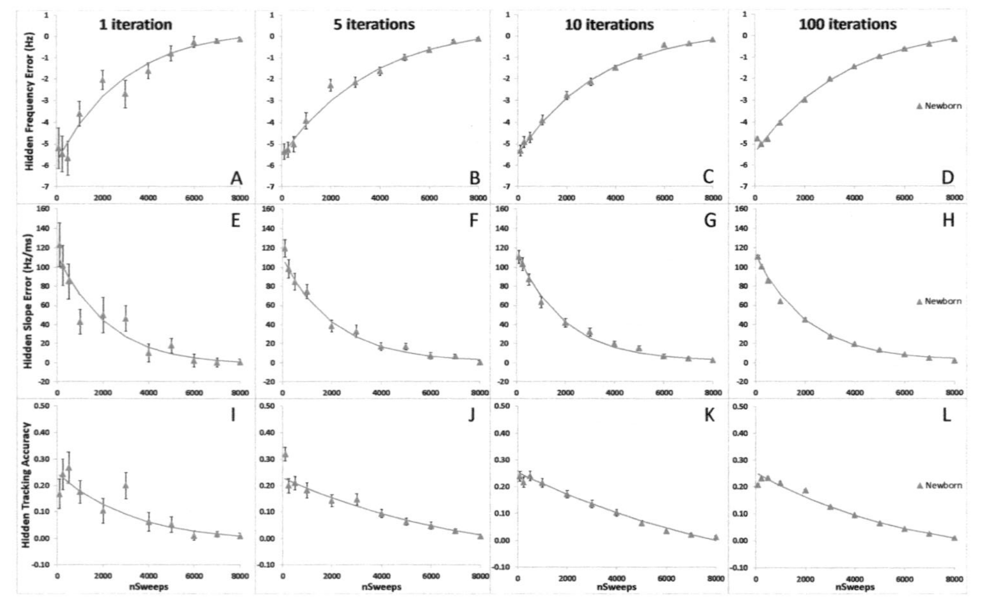

One-way ANOVA

1 Iteration.

For hFE, statistical results of the one-way ANOVA at 1 iteration demonstrated a significant difference in the nSweeps condition (F = 16.262, $p < 0.001$, $\eta_p^2 = 0.359$) factor. Results of all pair-wise comparisons are summarized in Appendix C.

For hSE, statistical results of the one-way ANOVA at 1 iteration demonstrated a significant difference in the nSweeps condition (F = 9.640, $p < 0.001$, $\eta_p^2 = 0.249$) factor. Results of all pair-wise comparisons are also summarized in Appendix C.

For hTA, statistical results of the one-way ANOVA at 1 iteration demonstrated a significant difference in the nSweeps condition (F = 5.694, $p < 0.001$, $\eta_p^2 = 0.164$) factor. Results of all pair-wise comparisons are also summarized in Appendix C.

5 Iterations.

For hFE, statistical results of the one-way ANOVA at 5 iterations demonstrated a significant difference in the nSweeps condition (F = 40.132, $p < 0.001$, $\eta_p^2 = 0.598$) factor. Results of all pair-wise comparisons are summarized in Appendix D.

For hTA, statistical results of the one-way ANOVA at 5 iterations demonstrated a significant difference in the nSweeps condition (F = 41.790, $p < 0.001$, $\eta_p^2 = 0.608$) factor. Results of all pair-wise comparisons are also summarized in Appendix D.

For hTA, statistical results of the one-way ANOVA at 5 iterations demonstrated a significant difference in the nSweeps condition (F = 11.389, $p < 0.001$, $\eta_p^2 = 0.297$) factor. Results of all pair-wise comparisons are also summarized in Appendix D.

10 Iterations.

For hFE, statistical results of the one-way ANOVA at 10 iterations demonstrated a significant difference in the nSweeps condition (F = 72.729, $p < 0.001$, $\eta_p^2 = 0.729$) factor. Results of all pair-wise comparisons are summarized in Appendix E.

For hSE, statistical results of the one-way ANOVA at 10 iterations demonstrated a significant difference in the nSweeps condition (F = 51.230, $p < 0.001$, $\eta_p^2 = 0.655$) factor. Results of all pair-wise comparisons are also summarized in Appendix E.

For hTA, statistical results of the one-way ANOVA at 10 iterations demonstrated a significant difference in the nSweeps condition (F = 23.824, $p < 0.001$, $\eta_p^2 = 0.469$) factor. Results of all pair-wise comparisons are also summarized in Appendix E.

100 Iterations.

For hFE, statistical results of the one-way ANOVA at 100 iterations demonstrated a significant difference in the nSweeps condition (F = 119.988, $p < 0.001$, $\eta_p^2 = 0.811$) factor. Results of all pair-wise comparisons are summarized in Appendix F.

For hSE, statistical results of the one-way ANOVA at 100 iterations demonstrated a significant difference in the nSweeps condition (F = 125.888, $p < 0.001$, $\eta_p^2 = 0.818$) factor. Results of all pair-wise comparisons are also summarized in Appendix F.

For hTA, statistical results of the one-way ANOVA at 100 iterations demonstrated a significant difference in the nSweeps condition (F = 45.414, $p < 0.001$, $\eta_p^2 = 0.619$) factor. Results of all pair-wise comparisons are also summarized in Appendix F.

Comparison of the Hidden Response Between Adults and Newborns (Aim 4b)

Finally, as a follow up to Aim 4a, the purpose of Aim 4b was to compare the adult and newborn hidden component of an FFR in an attempt to better understand the differences, if any, between the two age groups (adults and neonates).

Figure 21 depicts the comparative results of the hidden components for adults and newborns. Panel A displays the adult and newborn hFE results at 1 iteration. For the adult hidden component, hFE values were -5.634 ± 0.801 Hz at 1 iteration for 100 sweeps. With an increasing number of sweeps, the hFE values increased and plateaued at approximately 4000 sweeps. For the newborn hidden component, similarities were present in the overall response trends as those observed in the adult hidden component, with two exceptions. One difference was that the hFE values were higher at 100 sweeps (-5.196 ± 0.935 Hz). Another difference was that, roughly 6000 sweeps were required to reach a plateau for newborns. As expected, a higher number of sweeps was required to reach a plateau in the newborns. When viewing data from 5, 10, and 100 iterations (Panels B, C, & D), results were consistent with those observed at 1 iteration.

Panel E displays the results for the adult and newborn hSE feature. When viewing the adult hidden data for 1 iteration, hSE values were largest (117.310 ± 19.600 Hz/ms) at 100 sweeps. As the number of sweeps increased, hSE values decreased became closer to zero Hz/ms. a plateau. For the adult hidden component, a plateau occurred at roughly 4000 sweeps. For the newborn hidden data, at 100 sweeps s hSE values were 122.903 ± 22.799 Hz/ms. As the number of sweeps decreased, roughly 5000-6000 sweeps were required to reach a plateau comparable to the adult hidden component. As expected, a higher number of sweeps was required to reach a plateau in the newborns. Furthermore, the data from 5, 10, and 100 iterations (Panels F, G, & H) were similar to the 1 iteration results.

Finally, the results of the adult and newborn hTA feature can be viewed in Panel I. For the adult hidden component, at 1 iteration, hTA values were largest (0.340 ± 0.050) at 100 sweeps. With an increasing number of sweeps, hTA values decreased and reached a plateau. For the newborn hidden component, similarities were present in overall response trends as those observed in the adult hidden component, with two exceptions. One difference was that the hTA values for newborns were higher at 100 sweeps (0.167 ± 0.057). Another difference was that roughly 6000 sweeps were needed to reach a newborn hTA plateau. As expected, a higher number of sweeps was required to reach a plateau in the newborns. When comparing the data from 5, 10, and 100 iterations (Panels J, K, & L), results remained consistent with 1 iteration.

Figure 21

Graphical displays of the raw data extracts for the adult and newborn participants hidden component at 1, 5, 10 and 100 iterations for Hidden Frequency Error (first row), Hidden Slope Error (second row), and Hidden Tracking Accuracy (third row).

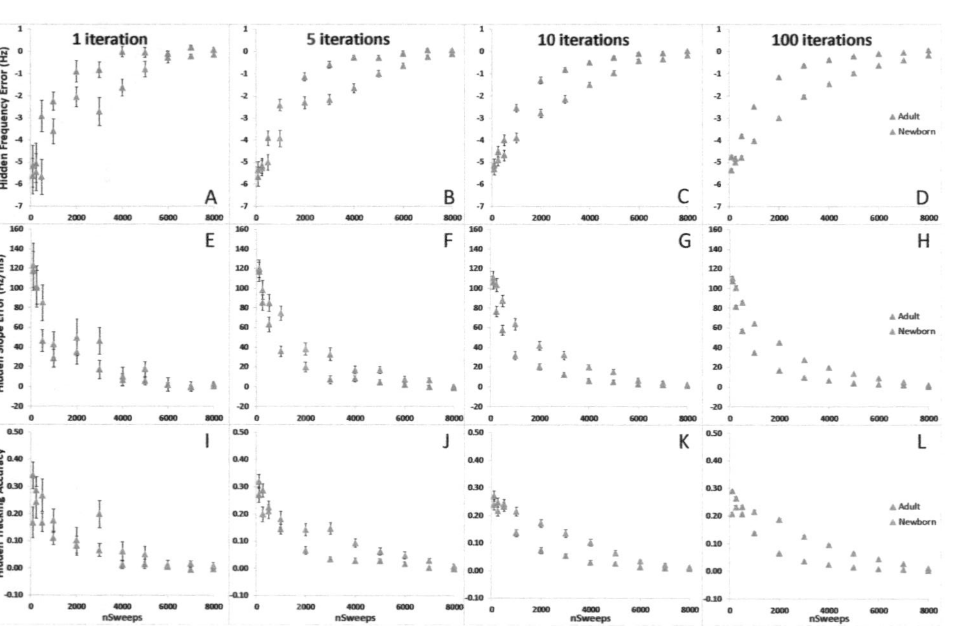

Linear Regression

Results of the linear regression analysis for the newborn hidden component can be seen in Figure 22. Similarly, to the previous linear regression analysis in this study Panel A depicts the FE results for the linear regression, Panel E depicts the SE results of for the linear regression, and Panel I depicts the TA results of for the linear regression. Equations for the linear regression can be viewed as follows.

1 Iteration.

$$\text{Adult Linear Regression for hFE: } y = 0.0006x + 3.647 \tag{199}$$

$$\text{Newborn Linear Regression for hFE } y = 0.0007x + 4.9126 \tag{200}$$

$$\text{Adult Linear Regression for hSE: } y = -0.0113x - 70.863 \tag{201}$$

$$\text{Newborn Linear Regression for hSE: } y = -0.0135x - 88.817 \tag{202}$$

$$\text{Adult Linear Regression for hTA: } y = -4\text{E-}05x - 0.2163 \tag{203}$$

$$\text{Newborn Linear Regression for hTA: } y = -3\text{E-}05x - 0.2182 \tag{204}$$

For hFE, Equations 199 and 200 depict the results for adult and newborn groups. The newborn slope was steeper than the adult slope. This meant that for the newborn groups hFE values increased at a faster rate compared to the adults. In other words, the increment for the newborn hFE values took a smaller number of sweeps to obtain a similar performance when compared to adults. When viewing the differing intercepts, values for the adult hidden component indicated a lower noise floor for the hFE trend.

hSE slope estimates can be viewed in Equations 201 and 202 for the adult and newborn hidden results. Slope estimates for hSE indicated a decremental trend for hSE as the number of sweeps increased. When comparing the adult and newborn hSE equations, there are observable differences between the slope estimates. It was noted that the newborn slope was steeper than that of the adult. Meaning that the number of sweeps needed for the newborn hSE to reach a comparable performance to the adult group was smaller. When viewing the different intercepts for the hidden component, values for the adult group indicated a lower noise floor for the hSE trend, when compared to those observed in the newborn group.

Finally, the adult and newborn hTA equations (203 & 204) estimated a decreasing trend in hTA as the number of sweeps increased. The adult and newborn slope estimates also displayed differing values. The newborn hTA slope was steeper than the adult slope. Simply put, the newborn group's hFE values decreased at a faster rate when compared to those observed in the adult group. In other words, the decrement for the adult hFE values took more time to obtain a similar performance as the newborn. When viewing the similar hTA intercepts, values for the adult and newborn groups indicated a comparable starting point for the hTA trend.

5 Iterations.

Adult Linear Regression for hFE: $y = 0.0007x - 3.9501$ (205)

Newborn Linear Regression for hFE $y = 0.0007x - 4.8006$ (206)

Adult Linear Regression for hSE: $y = -0.0114x + 69.198$ (207)

Newborn Linear Regression for hSE: $y = -0.0134x + 90.133$ (208)

Adult Linear Regression for hTA: $y = -3\text{E-}05x + 0.2129$ (209)

Newborn Linear Regression for hTA: $y = -3\text{E-}05x + 0.2331$ (210)

When viewing the 5 iteration equations, the values and trends remained similar to the 1 iteration results.

10 Iterations.

Adult Linear Regression for hFE: y = 0.0006x - 3.7854 (211)

Newborn Linear Regression for hFE y = 0.0007x - 4.6985 (212)

Adult Linear Regression for hSE: y = -0.0101x + 63.132 (213)

Newborn Linear Regression for hSE: y = -0.0131x + 88.188 (214)

Adult Linear Regression for hTA: y = -3E-05x + 0.2056 (215)

Newborn Linear Regression for hTA: y = -3E-05x + 0.2356 (216)

When viewing the 10 iteration equations, the values and trends remained similar to the 1 and 5 iteration results.

100 Iterations.

Adult Linear Regression for hFE: y = 0.0006x - 3.797 (217)

Newborn Linear Regression for hFE y = 0.0006x - 4.6498 (218)

Adult Linear Regression for hSE: y = -0.0104x + 64.092 (219)

Newborn Linear Regression for hSE: y = -0.0129x + 87.268 (220)

Adult Linear Regression for hTA: y = -3E-05x + 0.2065 (221)

Newborn Linear Regression for hTA: y = -3E-05x + 0.2311 (222)

When viewing the 100 iteration equations, the values and trends remained similar to the 1, 5, and 10 iteration results.

Figure 22

*Graphical presentation: of the linear regressions performed on the adult and newborn hidden component at 1, 5, 10 and 100
iterations for Hidden Frequency Error (first row), Hidden Slope Error (second row), and Hidden Tracking Accuracy (third row).*

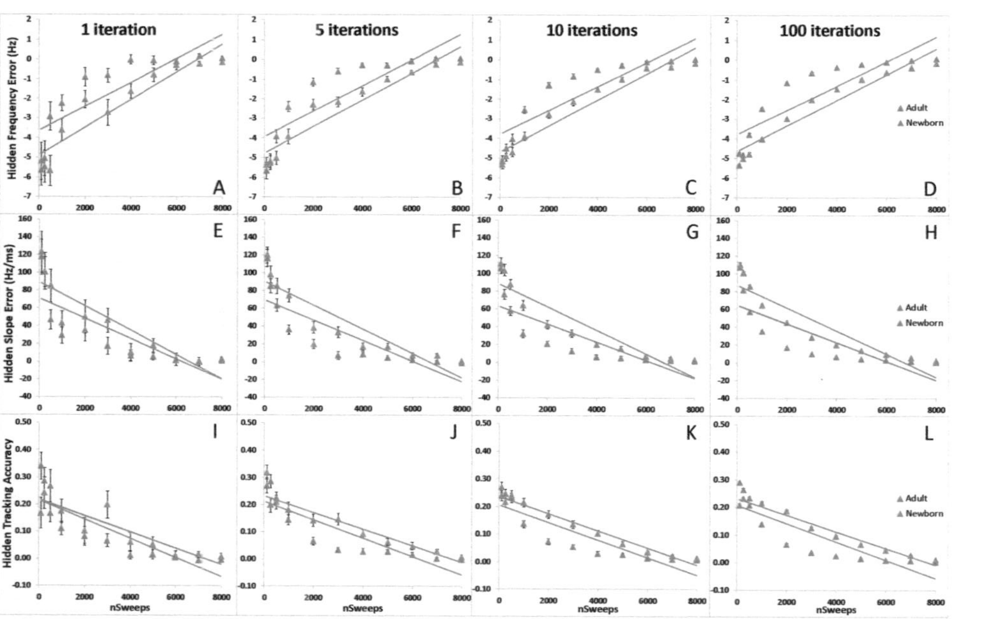

Exponential Modeling

Results of exponential modeling for the hidden component can be seen in Figure 23. In line with the previous regression figures, Panel A depicts the FE results for the exponential modeling, Panel E depicts the SE results for the exponential modeling, and Panel I depicts the TA results for the exponential modeling. Each row in the figure displays results obtained at 1, 5, 10, and 100 iterations, respectively. Exponential modeling equations are listed as follows. A tabular depiction of the Tau, A_{noise}, and A_{as} values can be viewed in Appendix G.

1 Iteration.

Adult Exponential Modeling for hFE: $A(n) = -0.015\left(1 - e^{-\frac{n}{972.560}}\right) + 5.634$ $\qquad$ (223)

Newborn Exponential Modeling for hFE: $A(n) = -0.061\left(1 - e^{-\frac{n}{2968.763}}\right) + 5.655$ $\qquad$ (224)

Adult Exponential Modeling for hSE: $A(n) = 117.310\left(e^{-\frac{n}{577.063}}\right) - 5.354$ $\qquad$ (225)

Newborn Exponential Modeling for hSE: $A(n) = 108.779\left(e^{-\frac{n}{2178.090}}\right) - 0.576$ $\qquad$ (226)

Adult Exponential Modeling for hTA: $A(n) = 0.012\left(e^{-\frac{n}{751.918}}\right) - 0.340$ $\qquad$ (227)

Newborn Exponential Modeling for hTA: $A(n) = 0.010\left(e^{-\frac{n}{3136.437}}\right) - 0.240$ $\qquad$ (228)

For hFE, the exponential modeling Equations 223 and 224 denoted differing values. The asymptotic amplitude (A_{AS}) for the adult and newborn groups were -0.015 Hz and -0.061 Hz, respectively. Similar hFE plateau values indicated that the age group had little to no influence on the hFE plateau. The τ values of the fitted curves for hFE trends obtained in the adult and newborn groups were 972 and 2968 sweeps, respectively. τ values showed that the adult participants produced a smaller τ value than neonates. This finding indicated that the adult participants displayed a faster improvement in hFE with increasing nSweeps when compared to the neonates. Meaning that, the hidden FE component plateaued earlier in adults than in the newborn group. The noise amplitude (A_{noise}) values for the adult and newborn groups were -5.634 Hz and -5.655 Hz, respectively. These values indicated that the A_{noise} estimates in the newborn group was 14% ([5.655 + 5.634] / 5.655 = 14%) lower than that of the adult group.

For hSE, the exponential modeling Equations (225 and 226) indicated a difference in values. The noise amplitude (A_{noise}) values for the adult and newborn groups were -117.310 Hz/ms and 108.779 Hz/ms, respectively. These values indicated that the A_{noise} estimates in the adult group was 7% ([117.310 – 108.779] / 117.310 = 7%) lower than that of the newborn group. The τ values of the fitted curves for hSE trends obtained in the adult and newborn groups were 577 and 2178 sweeps, respectively. This finding indicated that the newborn participants displayed a slower improvement on hSE with increasing nSweeps when compared to those obtained in the adult group. The asymptotic amplitude (A_{AS}) for the adult and newborn groups were 5.354 Hz/ms and 0.576 Hz/ms, respectively. These differing hSE plateau values indicated that the newborn age group had shown an influence on the hSE plateau values.

For hTA, the exponential modeling Equations 227 and 228 showed varying values. The noise amplitude (A_{noise}) values for the adult and newborn groups were 0.012 and 0.010, respectively. These values indicated that the A_{noise} estimates in the adult group was 17% ([0.012 - 0.010] / 0.012 = 17%) lower than that of the newborn group. The τ values of the fitted curves for hTA trends obtained in the adult and newborn groups were 751 and 3136 sweeps, respectively. This finding indicated that the adult participants displayed a faster improvement in hTA with increasing nSweeps when compared to the newborns. The asymptotic amplitude (A$_{AS}$) for the adult and newborn groups were 0.340 and 0.240, respectively. Slightly differing hTA plateau values indicated that the age group had shown an influence on the hTA plateau values.

5 Iterations.

Adult Exponential Modeling for hFE: $A(n) = -0.064\left(1 - e^{-\frac{n}{1153.987}}\right) + 5.666$ $\quad$ (229)

Newborn Exponential Modeling for hFE: $A(n) = -0.123\left(1 - e^{-\frac{n}{3783.525}}\right) + 5.376$ $\quad$ (230)

Adult Exponential Modeling for hSE: $A(n) = 102.178\left(e^{-\frac{n}{795.862}}\right) - 4.224$ $\quad$ (231)

Newborn Exponential Modeling for hSE: $A(n) = 105.366\left(e^{-\frac{n}{2055.145}}\right) - 3.415$ $\quad$ (232)

Adult Exponential Modeling for hTA: $A(n) = 0.012\left(e^{-\frac{n}{1106.627}}\right) - 0.317$ $\quad$ (233)

Newborn Exponential Modeling for hTA: $A(n) = 0.013\left(e^{-\frac{n}{7645.131}}\right) - 0.225$ $\quad$ (234)

When viewing the 5 iteration equations, the values remained similar to the 1 iteration results.

10 Iterations.

Adult Exponential Modeling for hFE: $A(n) = -0.057\left(1 - e^{-\frac{n}{1352.667}}\right) + 5.147$ $\quad$ (235)

Newborn Exponential Modeling for hFE: $A(n) = -0.177\left(1 - e^{-\frac{n}{3460.903}}\right) + 5.310$ $\quad$ (236)

Adult Exponential Modeling for hSE: $A(n) = 95.312\left(e^{-\frac{n}{742.458}}\right) - 4.849$ $\quad$ (237)

Newborn Exponential Modeling for hSE: $A(n) = 110.592\left(e^{-\frac{n}{1884.749}}\right) - 3.797$ $\quad$ (238)

Adult Exponential Modeling for hTA: $A(n) = 0.009\left(e^{-\frac{n}{1577.897}}\right) - 0.269$ $\quad$ (239)

Newborn Exponential Modeling for hTA: $A(n) = 0.001\left(e^{-\frac{n}{10667.208}}\right) - 0.248$ $\quad$ (240)

When viewing the 10 iteration equations, the values remained similar to the 1 and 5 iteration results.

100 Iterations.

Adult Exponential Modeling for hFE: $A(n) = -0.073\left(1 - e^{-\frac{n}{1194.074}}\right) + 5.368$ (241)

Newborn Exponential Modeling for hFE: $A(n) = -0.161\left(1 - e^{-\frac{n}{3588.506}}\right) + 5.252$ (242)

Adult Exponential Modeling for hSE: $A(n) = 96.525\left(e^{-\frac{n}{823.570}}\right) - 3.554$ (243)

Newborn Exponential Modeling for hSE: $A(n) = 107.736\left(e^{-\frac{n}{2078.514}}\right) - 4.515$ (244)

Adult Exponential Modeling for hTA: $A(n) = 0.007\left(e^{-\frac{n}{1212.950}}\right) - 0.289$ (245)

Newborn Exponential Modeling for hTA: $A(n) = 0.011\left(e^{-\frac{n}{6445.364}}\right) - 0.248$ (246)

When viewing the 100 iteration equations, the values remained similar to the 1, 5, and 10 iteration results.

Figure 23

Graphical displays of the exponential modeling performed on the adult and newborn hidden components at 1, 5, 10 and 100 iterations for Hidden Frequency Error (first row), Hidden Slope Error (second row), and Hidden Tracking Accuracy (third row).

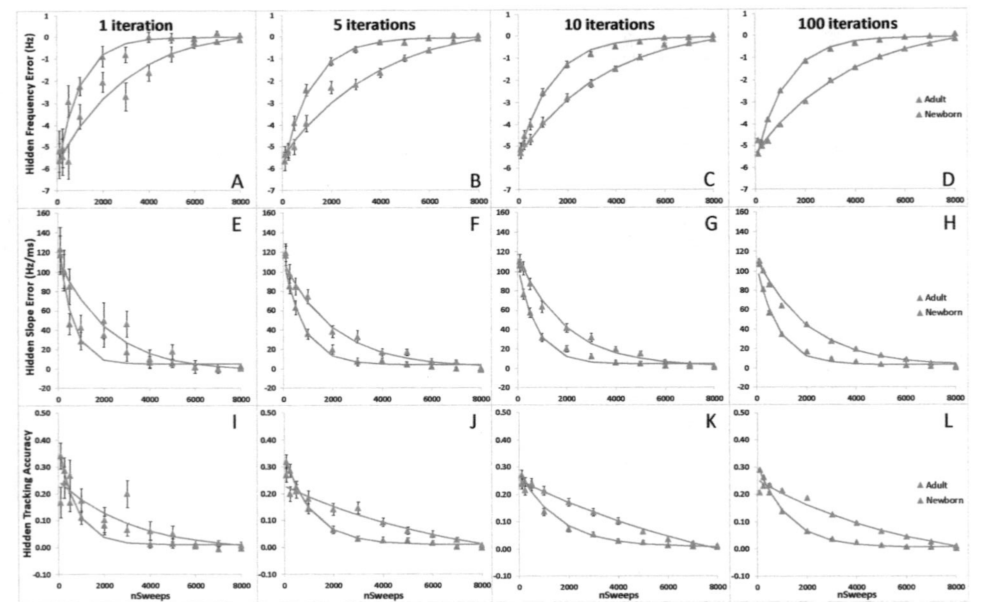

Two-way ANOVA

1 Iteration.

For hFE, statistical results of the two-way ANOVA at 1 iteration demonstrated a significant difference for the nSweeps condition ($F = 31.447$, $p < 0.001$, $\eta_p^2 = 0.364$), as well as the interaction between the age group ($F = 10.054$, $p = 0.002$, $\eta_p^2 = 0.155$) factors. For the age factor, hFE values for adults were significantly smaller (mean difference = 1.019 ± 0.321 Hz) than those obtained in the newborn group. Furthermore, the results of all pair-wise comparisons across the 11 nSweeps conditions are summarized in Appendix C.

For hSE, statistical results of the two-way ANOVA at 1 iteration demonstrated a significant difference for the nSweeps condition ($F = 20.206$, $p < 0.001$, $\eta_p^2 = 0.269$). The results for the hSE interaction between the age groups were nonsignificant ($F = 3.217$, $p = 0.078$, $\eta_p^2 = 0.055$). For the age factor, hSE values for adults were significantly larger (mean difference = -12.852 ± 7.165 Hz/ms than those obtained in the newborn group. Finally, the results of all pair-wise comparisons across the 11 nSweeps conditions are also summarized in Appendix C.

In regard to hTA, statistical results of the two-way ANOVA at 1 iteration demonstrated a significant difference for the nSweeps condition ($F = 14.792$ $p < 0.001$, $\eta_p^2 = 0.212$). The results for the hSE interaction between the age groups were nonsignificant ($F = 1.509$, $p = 0.224$, $\eta_p^2 = 0.027$). For the age factor, hTA values for adults were significantly larger (mean difference= -0.027 ± 0.02) than those obtained in the newborn group. For clarity, results of all pair-wise comparisons across the 11 nSweeps conditions are also summarized in Appendix C.

5 Iterations.

In regard to hFE, statistical results of the two-way ANOVA at 5 iterations demonstrated a significant difference for the nSweeps condition ($F = 85.493$, $p < 0.001$, $\eta_p^2 = 0.609$) as well as the interaction between the age groups ($F = 9.228$, $p = 0.004$, $\eta_p^2 = 0.144$). For the age factor, hFE values for adults were significantly smaller (mean difference $= 0.810 \pm 0.266$ Hz) than those obtained in the newborn group. Furthermore, the results of all pair-wise comparisons across the 11 nSweeps conditions are summarized in Appendix D.

With regard to hSE, statistical results of the two-way ANOVA at 5 iterations demonstrated a significant difference for the nSweeps condition ($F = 88.451$, $p < 0.001$, $\eta_p^2 = 0.617$) as well as the interaction between the age groups ($F = 6.375$, $p = 0.014$, $\eta_p^2 = 0.104$). For the age factor, hSE values for adults were significantly larger (mean difference $= -15.518 \pm 6.146$ Hz/ms) than those obtained in the newborn group. Finally, the results of all pair-wise comparisons across the 11 nSweeps conditions are also summarized in Appendix D.

For hTA, statistical results of the two-way ANOVA at 5 iterations demonstrated a significant difference for the nSweeps condition ($F = 30.825$, $p < 0.001$, $\eta_p^2 = 0.359$). The results for the hTA interaction between the age groups were nonsignificant ($F = 3.193$, $p = 0.079$, $\eta_p^2 = 0.055$). For the age factor, hTA values for adults were significantly larger (mean difference $= -0.032 \pm 0.018$ Hz) than those obtained in the newborn group. For clarity, results of all pair-wise comparisons across the 11 nSweeps conditions are also summarized in Appendix D.

10 Iterations.

For hFE, statistical results of the two-way ANOVA at 10 iterations demonstrated a significant difference for the nSweeps condition ($F = 101.603$, $p < 0.001$, $\eta_p^2 = 0.661$) as well as the interaction between the age groups ($F = 9.928$, $p = 0.003$, $\eta_p^2 = 0.160$). For the age factor, hFE values for adults were significantly smaller (mean difference = 0.791 ± 0.251 Hz) than those obtained in the newborn group. Furthermore, the results of all pair-wise comparisons across the 11 nSweeps conditions are summarized in Appendix E.

In regard to hSE, statistical results of the two-way ANOVA at 10 iterations demonstrated a significant difference for the nSweeps condition ($F = 94.313$, $p < 0.001$, $\eta_p^2 = 0.645$) as well as the interaction between the age groups ($F = 6.497$, $p = 0.014$, $\eta_p^2 = 0.111$). For the age factor, hSE values for adults were significantly larger (mean difference = -16.240 ± 6.372 Hz/ms) than those obtained in the newborn group. For clarity, results of all pair-wise comparisons across the 11 nSweeps conditions are also summarized in Appendix E.

For hTA, statistical results of the two-way ANOVA at 10 iterations demonstrated a significant difference for the nSweeps condition ($F = 45.466$, $p < 0.001$, $\eta_p^2 = 0.466$) as well as the interaction between the age groups ($F = 4.278$, $p = 0.044$, $\eta_p^2 = 0.076$). For the age factor, hTA values for adults were significantly smaller (mean difference = -0.034 ± 0.017) than those obtained in the newborn group. Finally, the results of all pair-wise comparisons across the 11 nSweeps conditions are also summarized in Appendix E.

100 Iterations.

In regard to hFE, statistical results of the two-way ANOVA at 100 iterations demonstrated a significant difference for the nSweeps condition ($F = 170.635$, $p < 0.001$,

$\eta_p^2 = 0.756$) as well as the interaction between age groups ($F = 10.948$, $p = 0.002$, $\eta_p^2 = 0.166$). For the age factor, hFE values for adults were significantly smaller (mean difference = 0.790 ± 0.239 Hz) than those obtained in the newborn group. Finally, the results of all pair-wise comparisons across the 11 nSweeps conditions are summarized in Appendix F.

When looking at hSE, statistical results of the two-way ANOVA at 100 iterations demonstrated a significant difference for the nSweeps condition ($F = 209.862$, $p < 0.001$, $\eta_p^2 = 0.792$) as well as the interaction between age groups ($F = 8.457$, $p = 0.005$, $\eta_p^2 = 0.133$). For the age factor, hSE values for adults were significantly larger (mean difference = -15.036 ± 5.170 Hz/ms) than those obtained in the newborn group. Furthermore, the results of all pair-wise comparisons across the 11 nSweeps conditions are also summarized in Appendix F.

For hTA, statistical results of the two-way ANOVA at 100 iterations demonstrated a significant difference in the nSweeps condition ($F = 89.031$, $p < 0.001$, $\eta_p^2 = 0.618$) as well as the interaction between age groups ($F = 6.421$, $p < 0.014$, $\eta_p^2 = 0.105$). For the age factor, hTA values for adults were significantly larger (mean difference = -0.037 ± 0.015) than those obtained in the newborn group. For clarity, results of all pair-wise comparisons across the 11 nSweeps conditions are also summarized in Appendix F.

Chapter 5: Discussion

This study sought to assess a potential algorithmic approach through non-negative matrix factorizations in attempts to minimize noise effects in FFR testing. Through a thorough investigation, optimal parameters were explored to assess noise reduction in FFR features. Results obtained from the present study indicated that the NMF algorithm could be a useful tool in the minimization of noise in FFR research. Furthermore, with the use of the algorithm, the number of sweeps needed to obtain an FFR could be minimized, therefore decreasing the overall recording time. With a substantial decrease in recording time, the FFR could be utilized outside of research scenarios (i.e., in clinical testing).

Understanding the Hidden Component of an FFR

The purpose of investigating the Hidden Component of an FFR was to present a novel approach in the extraction of the "hidden" portion of neural responses, in addition to the "apparent" portion of neural responses. Therefore, efforts were conducted to reduce the noise in the recording to a point where the "hidden" response was apparent. At earlier sweep counts, the adult and newborn results showed a large amount of noise separation. With the NMF algorithm, the separated noises were removed from the original recording and therefore could result in the assessment of the hidden component (Figures 21, 22, & 23). Findings from the hidden component indicated, that a response in FFR features were present at a lower number of sweeps but could not be visualized due to the dominating noise in the recording. Once the NMF algorithm was implemented, the hidden FFR features were revealed. This finding aided in the rationalization that the NMF algorithm could aid in overcoming the technical pitfalls associated with traditional signal averaging.

When viewing the adult hidden results, findings indicated that hidden component of the FFR was most apparent at lower sweep counts. From there, the results decreased and plateaued around ~5000-8000 sweeps (Figure 9). In other words, the greatest portion of uncovered responses for adults were in the range of 100-4000 sweeps. Through the use of the NMF algorithm, it was apparent that neural responses were present at lower sweep counts but could not be visualized through conventional signal averaging. With the NMF algorithm these responses were enhanced and visualized. The adult linear regression analysis indicated the overall trend of the hidden response (i.e., hFE had an incremental trend, hSE had a decremental trend, and also hTA had a decremental trend). Furthermore, exponential modeling revealed the timing aspects of the hidden component as well as the noise floor and plateau. These values will be further discussed after the next paragraph, in comparison to the newborn hidden results.

When reviewing the newborn hidden results, findings indicated that hidden component of the FFR recording was also most apparent at lower sweep counts. A visual plateau occurred around ~6000-8000 sweeps (Figure 18). This significance indicated that hidden responses were present at lower sweep counts but could not be visualized through conventional signal averaging alone, therefore needing the NMF algorithm to bring out such responses. Furthermore, this finding also indicated that the NMF algorithm does work on newborn recordings. Overall linear regression and exponential modeling results depicted similar trends when compared to the adult results

In comparing the adult and newborn hidden components, as mentioned, their overall trends were similar. Some interesting findings came from the values observed in the linear regression and exponential modeling. Specifically, for the linear regression analysis, the mean slopes for the hidden component of the adult and newborn groups were similar across each hidden FFR feature. The difference in the linear regression analysis occurred between the mean intercept values for each hidden FFR feature. For hFE, the adult mean intercept was -3.795 and the newborn was -4.762. For hSE, the adult mean intercept was 66.821 and the newborn was 88.602. For hTA, the adult mean intercept was 0.210 and the newborn was 0.230. Differing intercept values indicate that the noise floor for the adult group was lower those of the newborn hidden group. This finding means that the estimated noise amplitude in the adult group was lower than the estimated noise amplitude in the newborn group, indicating that the adults have a smaller SNR than the newborns. This can indicate that the newborn group has more noise in their recordings.

When comparing noise levels between age groups, results are consistent with Jeng and colleagues (2018). Their study conducted an in-depth investigation into the differences and similarities between newborns and adults while utilizing conventional signal averaging. Findings from their study revealed that although exponential modeling can be beneficial in the depiction of newborn trends, the high signal to noise ratios in the newborn results display less favorable outcomes when compared to the results of the adults. Furthermore, the newborn group requires a larger number of sweeps in order to average out and minimize the effects of noise in their recordings when compared to adult recordings. This finding is consistent throughout the literatures, as well as this current study. When the NMF algorithm is applied, the hidden portion of the responses revealed by the algorithm allows further exploration of the nature of the FFRs that are recorded from adults and neonates in two different testing environments.

When comparing the values for the hidden components exponential modeling, similarities were present as well as some differences between the adult and newborn groups. The noise floors and asymptotic amplitudes from each condition were comparable between the adult and newborn hidden components. For hFE, the adult and newborn values for A_{noise} were -5.45 Hz and -5.40 Hz and values for A_{AS} were -0.05 Hz and -0.13Hz, respectively. For hSE, the adult and newborn values for A_{noise} were 102.83 Hz/ms and 108.12 Hz/ms and values for A_{AS} were 4.49 Hz/ms and 3.07 Hz/ms, respectively. For hTA, the adult and newborn values for A_{noise} were both 0.01 and values for A_{AS} were 0.30 and 0.24, respectively. One difference is that the adult group displayed smaller tau values when compared to the newborn group (Figure 23). This finding indicates that the adult hidden component improved faster with increasing numbers of sweeps when compared to the newborn hidden component (adult and newborn mean tau hFE values 1168 & 3450 sweeps, hSE values 734 & 2048 sweeps, and hTA 1161 & 7000 sweeps, respectively).

The most interesting part about the exponential modeling results are the beginning and ending data points for each group. These values were nearly identical. Despite the similarities in starting and ending values, the NMF algorithm worked better on discovering the adult hidden component than the newborn hidden component. Reasons as to why this may have occurred are unknown but leads the researchers to wonder if it is attributed to the noisy nature of newborns, the neural immaturity of newborns, or the different recording environments.

Differences Between the ORI and NMF Conditions

When the NMF algorithm was applied to both the adult and newborn data, a consistent finding was observed in the overall enhancements of FFR features within the NMF condition. Simply put, when the NMF algorithm was implemented, the FFR features started at a better response level at 100 sweeps when compared to the ORI condition (Figures 6 & 12). This change indicated an overall baseline improvement across FFR features. With an increasing number of sweeps, the NMF condition continued to improve, but the rate was not as fast when compared to the ORI condition. This difference can be attributed to this overall enhancement from the NMF algorithm. This initial enhancement in features also minimized the complications observed at 100 and 250 nSweeps where the ORI condition displayed outliers which potentially skewed the data. The minimization of potential outliers, with the NMF algorithm, implies that the general variability among the lower sweep counts could be minimized. Simply put, the NMF algorithm worked effectively to extract targeted FFRs and in such removed a substantial amount of noise. Such improvements could be useful in the elimination of potential outliers.

When reviewing results from the adult ORI and NMF testing (Figure 6), with the use of the NMF algorithm, as little as 1000-4000 sweeps could be needed to visualize a plateau in FFR features. This finding was consistent with the work reported by Jeng and colleagues (unpublished data) that utilized similar protocols in noise separation through NMF algorithm. The largest improvements in the feature responses were observed at lower nSweeps counts (i.e., 100-4000). As the number of sweeps began to rise past this plateau, the FFR features showed no significant change in values, meaning that the differences between the 5000-8000 sweep range (i.e., > 4000 sweeps) were nonsignificant when compared between the ORI and NMF conditions. This finding indicated that the results of NMF algorithm at a higher number of sweeps were nonsignificant. This finding is consistent with the principle of signal averaging (Hyde, 1984). Which states that, with an accumulating number of sweeps, the noise in a recording can be minimized and averaged out. With a smaller number of sweeps (i.e., 100-4000), the NMF algorithm showed a significant benefit in the FFR feature values.

When viewing the results of the adult linear regressions, the ORI condition showed a steeper slope for each FFR feature (ORI and NMF mean slope values: FE -0.001 & -0.0004, SE 0.013 & 0.005, and TA 5E-05 & 2E-05, respectively). The rate of change for the feature was heavily dependent upon the number of sweeps in the recording for the ORI condition. This difference can be attributed to the overall enhancement of FFR features due to the NMF algorithm. As mentioned, with the use of the NMF algorithm, the initial FFR features were smaller. The intercepts for the ORI condition were generally larger than the intercepts for the NMF condition (ORI and NMF mean intercept values: FE 9.961 & 6.371, SE -139.700 & -27.141, and TA 0.56 & 0.77, respectively). This finding indicates that the noise floor in the ORI condition is much higher than the NMF. This verifies that the NMF algorithm is properly working to minimize the noises which are present in the ORI condition. The noise reductions in the ORI condition, through the smaller intercept values obtained in the linear regressions, are less apparent than those in the NMF condition.

For exponential modeling, in adults, the ORI and NMF conditions displayed differing values for the A_{noise} and tau values. Predominately the noise floor in the NMF condition was lower than the noise floor in the ORI condition (ORI and NMF mean A_{noise} values: FE 12.782 Hz & 7.715 Hz, SE -206.552 Hz/ms & -96.349 Hz/ms, and TA 0.42 & 0.66, respectively). Meaning that, the FFR feature values showed a better performance when the NMF algorithm was applied when compared to those for the ORI condition. Another difference was noted in the tau values for the ORI and NMF conditions. Information about the time constants can aid in the determination of the required number of sweeps needed in a recording. Specifically, the ORI tau values were predominately smaller at one iteration when compared to the NMF tau values at that same iteration. This finding indicated that the ORI condition showed a faster improvement in FFR at one iteration with increasing number of sweeps compared to those obtained by the NMF condition. These differences at one iteration can be attributed to the overall enhancement of the features shown by the NMF algorithm as well as the potential outliers at 100 and 250 sweeps in the ORI condition. At 5, 10 and 100 iterations, an opposite effect is seen. That is, the overall tau values are smaller in the NMF condition when compared to the ORI condition (ORI and NMF mean tau values: FE 922 & 720, SE 560 & 555, and TA 861 & 420 sweeps, respectively). The asymptotic amplitude (A_{As}) values were similar between the ORI and NMF conditions, indicating that the NMF algorithm had little to no influences on the plateau of the FFR features. Overall, exponential modeling provided greater detail into the differences between the ORI and NMF conditions.

When the NMF algorithm was applied to the newborn group, the recordings exhibited an overall enhancement in FFR features when compared to those obtained in the ORI condition (Figure 12). For neonates, a minimum number of 2000-6000 sweeps were needed to observe a plateau in FFR features when the NMF algorithm was applied.

This reduction in nSweeps (i.e., <6000-8000), when the NMF algorithm was utilized, could indicate a time reduction in overall FFR testing. As previously stated, in newborn testing, timing is a key aspect to consider. Through the use of the NMF algorithm, testing time could be shortened. This could aid in the implementation of the FFR in clinical testing scenarios.

Similar to the overall trends in adults, when viewing the results of the newborn linear regressions, the ORI condition showed a steeper slope for each FFR feature (ORI and NMF mean slope values: FE -0.012 & -0.001, SE 0.021 & 0.007 and TA 6.8E-05 & 4E-05, respectively). This can be attributed to the overall enhancement of FFR features due to the NMF algorithm as well as the potential outliers in the ORI condition at 100 and 250 sweeps. When viewing the intercept values, the NMF values were smaller than the values obtained in the ORI condition (ORI and NMF mean intercept values: FE 12.322 &7.072, SE -192.348 & -96.218, and TA 0.406 & 0.636, respectively). As previously mentioned for the adult intercepts, this similar finding indicates that the noise floor in the NMF condition is much lower than that in the ORI condition. This means that the NMF algorithm is working effectively in minimizing the amount of noise which is still present in the ORI condition.

For the newborn exponential modeling, the NMF and ORI conditions were consistent across iterations. Overall, the noise floor in the NMF condition was lower than the noise floor for the ORI condition (ORI and NMF mean A_{noise} values: FE 13.571 Hz & 8.469 Hz, SE -233.021 Hz/ms & -122.744 Hz/ms, and TA 0.34 & 0.59, respectively). The ORI tau values were predominately larger when compared to the NMF values (ORI and NMF mean tau values: FE 2641 & 1932, SE 1927 & 1919 and TA 3204 & 1992 sweeps, respectively). The A_{AS} between the two conditions showed similar values, indicating that there were no influences on the feature plateau. This means that the plateau for the A_{AS} values were not dependent upon the NMF or ORI conditions. Results from the newborn exponential modeling indicate that the general noise in the NMF recording is consistently smaller than the noise in the ORI recording. The tau values aid in the depiction of the of the time it would take for each condition to reach 63% of the asymptotic amplitude. The results of the tau values show that the NMF condition displays a substantial time decrement than those obtained or the ORI condition. General implications of these findings show that the NMF algorithm is not only working to minimize the noise in the FFR, but also can aid in the time reduction for the recording.

Similarities and Differences Between the Adult and Newborn Groups

When investigating the spectrograms before and after the application of the NMF algorithm, consistent results were noted in regard to the overall enhancement and trend of the FFR features in adults and newborns. However, differences were also observed. One difference occurred after the NMF algorithm was applied to the newborns data. Interestingly enough, the newborn group displayed a larger variability across each feature when compared to the adult group, even after the application of the NMF algorithm (Figure 15). Differences between the two age groups were also observed in the FFR feature plateaus. For newborn recordings, a higher number of sweeps (2000-6000) was required to reach a plateau. For the adults, roughly 1000-4000 sweeps were needed to observe a plateau in FFR features. The larger number of sweeps needed for the newborn group could be attributed to the neural immaturity in newborns, as well as the overall noisy nature of newborns and their recording environments (Jeng et al., 2018). This finding was consistent with initial predictions that the nSweeps plateau would occur at a higher number of sweeps for the newborn group than those observed for the adult group. Despite the higher feature plateaus for newborns, there was still a substantial reduction with the NMF algorithm when compared to the ORI condition. This finding means that the NMF algorithm could provide a feasible method for the analysis of FFR testing across ages.

When comparing the linear regression results, similarities were observed in the overall trends for the adult and newborn groups. For FE a decremental trend was observed and for SE and TA an incremental trend was observed. The slope and intercept values were also similar when compared between the two groups. For FE, the adult and newborn slopes were -0.0004 and -0.0005, and the intercept values were 6.371 and 7.072, respectively. For SE, the adult and newborn slopes were 0.005 and 0.006, and the intercept values were -80.407 and -96.217, respectively. For TA, the adult and newborn slopes were 2E-05 and 3E-05, and the intercept values were 0.7 and 0.6, respectively.

The exponential modeling again shows similar trends in overall responses (Figure 17) between the adult and newborn groups, but also differ in a few ways. The noise floors and asymptotic amplitudes from each condition were relatively comparable between the adult and newborn NMF groups. For FE, the adult and newborn values for A_{noise} were 7.714 Hz and 8.468 Hz and values for A_{AS} were 3.655 Hz and 4.148 Hz, respectively. For SE, the adult and newborn values for A_{noise} were -96.349 Hz/ms and -122.744 Hz/ms and values for A_{AS} were -46.583 Hz/ms and -59.647 Hz/ms, respectively. For TA, the adult and newborn values for A_{noise} were 0.669 and 0.586 and the A_{AS} values were 0.88 and 0.87, respectively. Differences between the models stemmed from the tau values. Overall tau values were smaller for the adult NMF group when compared to the newborn NMF group (adult and newborn NMF mean tau values: FE 720 & 1932, SE 555 & 1919 and TA 420 & 1992 sweeps, respectively). These findings are similar to the results of the hidden component and indicates that, although the NMF algorithm shows great improvements on both adult and newborn FFRs, the algorithm works slightly better on adult recordings.

Optimal Parameters for Utilizing the NMF Algorithm in FFR Testing

Through a thorough evaluation of the NMF algorithm, optimal recording parameters were assessed. When observing the effects of noise separation through the NMF algorithm on the raw data, it was apparent that the NMF algorithm showed a great improvement in spectral discernibility at a lower number of sweeps. Also, the separated noise was most apparent at a lower number of sweeps when the NMF algorithm was applied. A neural response was visually identifiable at earlier sweep counts for both adult and newborn recordings. With the use of the NMF algorithm, an enhancement of the FFR at a lower sweep counts, could ideally indicate a potential reduction in recording time.

When discussing the optimal parameters for testing, the most important factor to establish is the nSweeps condition. As previously mentioned, the number of sweeps used in FFR testing with conventional signal averaging can range from 1000 to 6000 sweeps (Skoe & Kraus, 2010), or even up to 8000 sweeps (Jeng et al., 2018). This need for a large number of sweeps lengthens the recording time. When FFR testing is used in conjunction with NMF algorithm, for adults, as few as 1000 sweeps are needed. When testing newborns, with the NMF algorithm as few as 2000 sweeps are needed. These sweep thresholds are based off of the tau values obtained in the adult and newborn exponential modeling for the NMF condition.

To provide an example of the time it would take to complete an adult FFR with the NMF algorithm, based off the tau value, for the American vowel /i/ with a rising frequency contour, it takes approximately 4.9 minutes to obtain 1000 accepted sweeps (i.e., (250 ms + 45 ms silent interval) x 1000 sweeps / 1000 = 590 seconds / 60 = 4.9 minutes). When compared to the original time of 39.3 minutes at 8000 sweeps with conventional signal averaging, this reduction is most ideal. For newborn testing with the NMF algorithm, it takes roughly 9.8 minutes to obtain 2000 accepted sweeps (i.e., (250 ms + 45 ms silent interval) x 2000 sweeps / 1000 = 590 seconds / 60 = 9.8 minutes). This sweep decrement, with the NMF algorithm, for both the adult and newborn groups can lead to an overall time decrease in a FFR recording session.

Based on the findings from exponential modeling, through the use of the NMF algorithm, the recording time for an FFR could potentially be cut in half. Meaning that the timeframe, originally at 40 minutes with conventional signal averaging, can be decreased to 5-10 minutes. This decrease could aid in FFR utility making this once lengthy testing tool more applicable in varying testing scenarios. One factor to still consider though is the age of the participant, researchers should anticipate a shorter recording time for adults than newborns.

An interesting result that should be noted was that the 100 iterations condition did not reveal the overall nIterations plateau as predicted. Based off the study conducted by Hart and Jeng (2021), it was predicted that when nIterations are manipulated, an increase in FFR features will be visualized with increasing nIterations and a stabilization should be observed around 100 iterations. Although overall variance was minimized at 100 iterations, this condition could be viewed as unnecessary for the purpose of this study. 100 iterations became procedurally time extensive and did not yield results that were significantly different. For this study, the more observable differences for the nIterations condition occurred somewhere between 5 and 10 iterations. Meaning that, the variability among FFR features were predominately minimized somewhere between 5 and 10 iterations. This finding was consistent with Llanos and colleagues (2017), where they achieved response stability utilizing 50 to 100 samples with a Hidden Markov Model.

Utilizing Machine Learning to Enhance Evoked Potentials

As previously mentioned, the NMF is a subset of machine learning. Pros to utilizing this algorithm are that, when compared with traditional algorithms, this method has shown ease of implementation, good interpretability of the results, and a small storage space (Yu et al., 2014). This method has also shown to be beneficial in the decomposition of multivariate data (Lee & Seung, 1999). With the use of the NMF algorithm researchers are able to separate noise effects from a recording. In a paper by Lin and colleagues (2017), recordings from various bat caves in national forests were obtained. The researchers utilized NMF procedures to tease out various noise components (i.e., road noise, wind noise, other animal sound) that were unrelated to the targeted sound source (i.e., bat calls). This study laid the foundation for NMF research, and thus provided the rationale for the implementation of the NMF algorithm in separating noise from FFRs.

The foundations of Lin and colleagues research led to this current study in assessing the NMF algorithm as an evaluator for noise separations in FFRs. As a result, this study investigated the integration of machine learning in modern EP testing. Previous research endeavors have begun to investigate the benefit of machine learning in EP testing (Dobrowolski et al., 2016; Hart & Jeng, 2018, 2020; Llanos et al., 2017; Xie et al., 2019). Limitations mentioned in previous studies were predominately based off the need for a large number of recorded sweeps in order to overcome the effects of noise in a recording (Hart & Jeng, 2018, 2020; Llanos, Xie, & Chandrasekaran, 2017; Xie et al., 2019).

For example, in the study conducted by Llanos and colleagues (2017) researchers investigated the potential integration of a Hidden Markov Model for the decoding and recognition of FFRs. The study sought to integrate machine learning techniques to overcome the technical issues such as poor SNRs in recordings. The results of their study show that machine learning can be utilized in the decoding of FFRs, but more investigations are required to minimized technical pitfalls related to the overall signal to noise ratio.

Another limitation mentioned in previous literatures were the uses of supervised machine learning in order to obtain classified results (Hart & Jeng, 2018, 2020; Llanos et al., 2017; Xie et al., 2019). Supervised machine learning requires operator influences (i.e., known inputs and outputs) in order for the algorithm to classify a response. Therefore, testing is not automated and independent, the need for human supervision is required. Another drawback when utilizing supervised learning is related to the assumptions that are made subjectively by the experimenters to indicate the presence or absence of a response in a recording (Xie et al., 2019). These operator influences can alter the outputs of an algorithm and show favorable results.

For example, Xie and colleagues (2019) utilized SVMs to classify FFRs in response to varying stimuli. Findings from their study indicated that the numbers of sweeps required for classifying FFRs in response to the four different Mandarin tones were much smaller those reported in the FFR and machine learning literature (Hart & Jeng, 2018, 2020). Although the SVM algorithm and other supervised machine learning algorithms may be useful in the classification of FFR recordings, the performances of those algorithms rely heavily on the experimenter's subjective judgments on the presence or absence of a response in a recording. To overcome these issues, the NMF algorithm provides an alternative approach that does not require any input from the experimenter's subjective judgments. As such, the NMF algorithm avoids those pitfalls that are commonly encountered in the realm of supervised machine learning.

Novelties of this study and optimal parameter search indicate that the integration of the NMF algorithm can provide a beneficial assessment of noise separation in FFRs. The NMF algorithm works as through unsupervised machine learning to accomplish feature extractions. With the NMF algorithm there is no need for operator influence or input output classifications. Through this integrated approach, researchers could learn more about the FFR analysis on a deeper level, based upon the results obtained from the novel extraction of hidden FFR features. These findings could provide significant information on pitch processing in individuals who cannot partake in traditional behavioral testing. A study conducted by Van Dyke and colleagues (2017) provided valuable research information in obtaining FFRs from infants ages 2-12 months filling in the age gap between newborns and adults. Knowing that FFRs can be obtained from any age group, such as immediate post-natal days (Jeng et al., 2018; Ribas-Prats et al., 2019) all the way into adulthood (Skoe & Kraus, 2010) the need for integration outside of research is apparent.

Study Limitations, Considerations, and Future Directions

This study sought to remove noise from FFR recordings while utilizing preexisting data. In the recordings chosen, a small silent interval (~45ms) was applied for each testing scenario. When using a small silent interval, recording time is decreased. This aids in a time reduction for typical testing scenarios, but when investigating noise effects, it can be less ideal. In any given recording, the silent interval is the portion of the response containing no stimulus information; therefore, the silent interval is essentially a compilation of noise throughout the recording. With a longer silent interval more noise is allowed into the recording. Therefore, the chance to study the "true noise" in the recording is greater. This could be utilized in conjunction with the NMF algorithm to aid in noise matching and separation. In future studies it may be beneficial to consider a lengthier silent interval for comparative measures in NMF noise separation.

Also, for this study maxSweeps was fixed at 8000 sweeps. This means that the fixed number of sweeps for the recording was set at 8000, and from there subsequent data analysis was conducted utilizing variations of the 8000 sweeps. In future studies it would be interesting to investigate varying cutoffs in maxSweeps values to understand the overall effect it could have on the NMF algorithm. In future investigations, the maxSweeps condition should be tested at 2000, 4000, and 6000 sweeps to gauge the accuracy of the NMF algorithm. Based on the FFR feature plateaus seen at as little as 1000 sweeps, perhaps a lower maxSweeps count may yield significantly different results.

Another consideration is to begin the assemblance of a FFR database. A system in which FFR data from other research collaborators can be stored and shared to diversify testing populations and widen research pool. Through this system the limitations of insufficient subject data may be overcome. This would eliminate the need for the iteration procedure and the potential biases which can potentially be introduced from the iteration technique (Hart & Jeng, 2020). As observed in this study, 100 iterations were excessive and unnecessary in the use of the NMF algorithm, but 5 and 10 iterations were helpful in minimizing the larger variations in the data. If a database was created for the access of unlimited participant data, the iteration procedure would no longer be implemented. This system would not only further benefit the overall study, but as well as benefit the FFR research community. Until there is a higher data count or the creation of a shared database, a further investigation should be conducted around 5-10 iterations to find the ideal participant number for algorithm training and testing.

Albeit the NMF approach has shown benefit in FFR enhancement among individuals with identifiable FFRs, there is still a need for an investigation of recordings from a disordered population. In order to fully establish the efficacy of the NMF algorithm in noise separations, procedures must be applied to FFR recordings where no visually identifiable FFR is present. To assess this method, one approach can be conducted from utilizing FFRs recordings obtained from normal hearing individuals which have been visually identified as absent. The NMF algorithm will then be implemented to separate any noise and reveal if there was true neural response in the recording.

Once this validation procedure is completed a further analysis can be conducted on a disordered population. Given that individuals with APD (Rocha-Muniz et al., 2012), SLI (Rocha-Muniz et al., 2012), dyslexia (Chandrasekaran et al., 2009; Ramus, 2012), and autism (Russo et al., 2008) have shown to have deficits in FFR recordings, obtaining data from individuals with these diagnoses would aid in subsequent data analysis. This would furthermore reveal how the NMF algorithm can separate any noise influences from the FFRs obtained from a disordered population.

Clinical Application

In the past, the only way to reliably interpret an EP test was by an experienced clinician or researcher. This individual would have gone through extensive training in the field to perform such tasks. Throughout the years, extensive research has led to the elimination of manual decision making in both ABR and ASSR recordings (Don, Eberling, & Waring, 1984; John & Picton, 2000). This automated procedure is independent from an observer's interpretation, and therefore, results in less technical training. If one were able to develop a similar method for the FFR, the need for operator judgements would be minimized. This accessibility would promote the further advancement of FFR testing in both clinical and research settings.

Through the development of an automated system, where FFR detection is generated though a computer, early assessment and detection of pitch processing in individuals could be achieved. This assessment could aid in the clinical evaluation of individuals who cannot provide a definitive behavioral response, such as infants, children, and individuals with a cognitive deficit. The crucial factor in this procedure is the minimization of noise effects in the recording, as well as an automated analysis.

Future studies need to be conducted in order to overcome the technical gaps needed to move forward towards FFR noise reduction and automation.

Chapter 6: Conclusions

In conclusion, findings reported in this study serve as a foundational stepping-stone in demonstrating the applicability of the NMF algorithm in noise separations for adult and newborn FFRs. The NMF algorithm has shown to be beneficial in the noise reduction as well as potential time reductions in FFR recordings, as seen in the linear regressions and exponential modeling. Furthermore, the overall enhancement of FFR features observed when the NMF algorithm was applied can indicate overall NMF effectiveness when compared to traditional signal averaging. The novel investigation of the hidden component revealed responses at lower sweep counts. These responses revealed by the algorithm enabled further exploration of the nature of the FFRs that are recorded from adults and neonates in two different testing environments.

Clinical and practical implications of this these findings include the use of the NMF algorithm in conjunction with FFR testing to boost clinical viability. This could furthermore increase FFR testing for screening and interventional services for individuals at risk for decreased pitch processing. Future studies should investigate integrative methods from this approach in order to take steps toward in FFR noise reductions and testing automation. Through the elimination of operator influence, an automated procedure could require less technical training for an individual to determine the presence or absence of an FFR. Through this future direction, the further development of the FFR could result in the advancement of this research tool into clinical applications.

References

Anderson, S., & Kraus, N. (2010). Sensory-Cognitive Interaction in the Neural Encoding of Speech in Noise: A Review. *Journal of the American Academy of Audiology, 21*(9), 575-585.

Anderson, S., Parbery-Clark, A., White-Schwoch, T., & Kraus, N. (2015). Development of subcortical speech representation in human infants. *Journal of the Acoustical Society of America 137*(6), 3346–3355.

Banai, K., Nicol, T., Zecker, S. G., & Kraus, N. (2005). Brainstem timing: Implications for cortical processing and literacy. *The Journal of Neuroscience: The Official Journal of the Society for Neuroscience, 25*(43), 9850–9857.

Banai, K., Abrams, D., & Kraus, N. (2007). Sensory-based learning disability: Insights from brainstem processing of speech sounds. *International Journal of Audiology, 46*(9), 524-532.

Ballachanda, B., & Moushegian, G. (2000). Frequency-following response: effects of interaural time and intensity differences. *Journal of The American Academy of Audiology, 11*(1), 1-11.

Carcagno, S., & Plack, C. J. (2011). Pitch discrimination learning: specificity for pitch and harmonic resolvability, and electrophysiological correlates. *Journal of the Association for Research in Otolaryngology: JARO, 12*(4), 503–517.

Chandrasekaran, B., Hornickel, J., Skoe, E., Nicol, T., & Kraus, N. (2009). Context-dependent encoding in the human auditory brainstem relates to hearing speech in noise: Implications for developmental dyslexia. *Neuron, 64*(3), 311–319.

Chandrasekaran, B., Krishnan, A., & Gandour, J. T. (2009). Relative influence of musical and linguistic experience on early cortical processing of pitch contours. *Brain and Language, 108* (1), 1–9.

Cherko, M., Hickson, L., & Bhutta, M. (2016). Auditory deprivation and health in the elderly. *Maturitas, 88,* 52–57.

Courant, R., Robbins, H., & Stewart, I. (1996). W*hat is mathematics? An elementary approach to ideas and methods.* Oxford University Press.

Delgado, R., & Ozdamar, O. (1994). Automated auditory brainstem response interpretation. IEEE *Engineering in Medicine and Biology Magazine, 13*(2), 227–237.

Dobrowolski, A., Suchocki, M., Tomczykiewicz, K., & Majda-Zdancewicz, E. (2016). Classification of auditory brainstem response using wavelet decomposition and SVM network. *Biocybernetics and Biomedical Engineering, 36,* 427–436.

Don, M., Elberling, C., & Waring, M., (1984). Objective detection of averaged auditory brainstem responses. *Scandinavian Audiology, 13,* 219-228.

Eimas, P. D., & Miller, J. L. (1992). Organization in the Perception of Speech by Young Infants. *Psychological Science, 3*(6), 340-345.

Elberling, C., & Don, M., (1984). Quality estimation of averaged auditory brainstem responses. *Scandinavian Audiology, 13,* 187- 197.

Elberling, C., & Don M., (1987). Detection functions for the human auditory brainstem response. *Scandinavian Audiology, 16,* 89-92.

Galbraith, G., Arbagey, P., Branski, R, Comerci, N., & Rector, P. (1995). Intelligible speech encoded in the human brain stem frequency- following response. *NeuroReport. 6,* 2363–2367.

Galbraith, G. (1994). Two-channel brain-stem frequency-following responses to pure tone and missing fundamental stimuli. *Electroencephalography and Clinical Neurophysiology. 92*, 321–330.

Galbraith, G. C., Bagasan, B., & Sulahian, J. (2001). Brainstem Frequency-following Response Recorded from One Vertical and Three Horizontal Electrode Derivations. *Perceptual and Motor Skills, 92*(1), **99-106.**

Gardi, J., Merzenich, M., & Mckean, C. (1979). Origins of the Scalp-Recorded Frequency-Following Response in the Cat. *International Journal of Audiology, 18*(5), 353–380.

Goldstein, L. J., Lay, D., Schneider, D. I., & Asmar, N. H. (2009). *Calculus & its applications*. Pearson.

Glaser E., Suter, C., Dasheiff, R., & Goldberg, A. (1976) The human frequency-following response: Its behavior during continuous tone and toneburst stimulation. *Electroencephalography and Clinical Neurophysiology. 40*, 25-32

Hart, B., & Jeng, F.-C. (2018). Machine learning in detecting frequency-following responses. *Proceedings of Meetings on Acoustics, 35*(1), 050002.

Hart, B., & Jeng, F.-C. (2020). Machine learning in detecting frequency-following responses in American neonates. [Unpublished manuscript]. Department of Communication Sciences and Disorders, Ohio University, Athens, Ohio

Hornickel, J., Knowles, E., & Kraus, N. (2012). Test-retest consistency of speech-evoked auditory brainstem responses in typically developing children. Hearing Research, 284(1-2), 52–58.

Hornickel, J., Skoe, E., Nicol, T., Zecker, S., & Kraus, N. (2009). Subcortical differentiation of stop consonants relates to reading and speech-in-noise perception. *Proceedings of the National Academy of Sciences, 106*(31), 13022-13027.

Hou, S. M., & Liscomb, D. M. (1979). An investigation of the auditory frequency-following responses as compared to cochlear potentials. *Archives of Otorhinolaryngology, 222*(3), 235-240.

Hyde, M. L. (1994). Signal processing and analysis. In J. T. Jacobson (Ed.), Principles and applications in auditory evoked potentials (pp. 47–83). Boston: Allyn & Bacon.

Jacobson, J. T., Jacobson, C.A., & Spahr, R. C. (1990) Automated and conventional ABR screening techniques in high-risk infants. *Journal of the American Academy of Audiology, 1*, 187-195.

Jeng, F., Hu, J., Dickman, B., Montgomery-Reagan, K., Tong, M., Wu, G., & Lin, C. (2011). Cross-linguistic Comparison of Frequency-following Responses to Voice Pitch in American and Chinese Neonates and Adults. Ear and Hearing, 32(6), 699–707.

Jeng, F.-C., Lin, C.-D., Chou, M.-S., Hollister, G. R., Sabol, J. T., Mayhugh, G. N., & Wang, C.-Y. (2016). Development of Subcortical Pitch Representation in Three-Month-Old Chinese Infants. Perceptual and Motor Skills, 122(1), 123–135.

Jeng, F.-C., Lin, C.-D., & Wang, T.-C. (2016). Subcortical neural representation to Mandarin pitch contours in American and Chinese newborns. The Journal of the Acoustical Society of America, 139(6), EL190.

Jeng, F.-C., Schnabel, E. A., Dickman, B. M., Hu, J., Li, X., Lin, C.-D., & Chung, H.-K. (2010). Early Maturation of Frequency-Following Responses to Voice Pitch in Infants with Normal Hearing. *Perceptual and Motor Skills, 111*(3), 765–784.

Jeng, F.-C., Nance, B., Montgomery-Reagan, K., & Lin, C.-D. (2018). Exponential modeling of frequency-following responses in American neonates and adults. *Journal of the American Academy of Audiology, 29*(2), 125-134.

Jeng, F.-C. (2020 unpublished data). Non-negative matrix factorization separates noise from frequency-following responses: A feasibility study.

John, M. S., Lins, O. G., Boucher, B. L., & Picton, T. W. (1998). Multiple Auditory Steady-state Responses (MASTER): Stimulus and Recording Parameters. *International Journal of Audiology, 37*(2), 59-82.

John, M., & Picton, T. (2000). MASTER: A Windows program for recording multiple auditory steady-state responses. *Computer Methods and Programs in Biomedicine, 61*(2), 125-150.

Johnson, K. L., Nicol, T. G., Zecker, S. G., & Kraus, N. (2007). Auditory brainstem correlates of perceptual timing deficits. *Journal of Cognitive Neuroscience, 19*(3), 376–385.

Kraus, N., & White-Schwoch, T. (2017). Neurobiology of Everyday Communication: What Have We Learned From Music? *The Neuroscientist, 23*(3), 287–298.

Krishnan, A. (2002). Human frequency-following responses: Representation of steady-state synthetic vowels. *Hearing Research, 166*(1-2), 192-201.

Krishnan, A., Xu, Y., Gandour, J. T., & Cariani, P. A. (2004). Human frequency-following response: Representation of pitch contours in Chinese tones. *Hearing Research, 189*(1-2), 1-12.

Krishnan, A., & Gandour, J. T. (2014). Language Experience Shapes Processing of Pitch Relevant Information in the Human Brainstem and Auditory Cortex: Electrophysiological Evidence. *Acoustics Australia, 42*(3), 166–178.

Kuhl, P. K. (1979). Speech perception in early infancy: Perceptual constancy for spectrally dissimilar vowel categories. *The Journal of the Acoustical Society of America, 66*(6), 1668-1679.

Kuhl, P. K. (2010). Brain mechanisms in early language acquisition. *Neuron, 67*, 713–727.

Kuhl, P. K., Stevens, E., Hayashi, A., Deguchi, T., Kiritani, S., & Iverson, P. (2006). Infants show a facilitation effect for native language phonetic perception between 6 and 12 months. *Developmental Science, 9*(2).

Kuhl, P., Williams, K., Lacerda, F., Stevens, K., & Lindblom, B. (1992). Linguistic experience alters phonetic perception in infants by 6 months of age. *Science, 255*(5044), 606-608.

Lee, D. D., & Seung, H. S. (1999). Learning the parts of objects by non-negative matrix factorization. *Nature, 401*(6755), 788–791.

Lin, F., Ferrucci, L., An, Y., Goh, J., Doshi, J., Metter, E., & Resnick, S. (2014). Association of hearing impairment with brain volume changes in older adults. *NeuroImage, 90*, 84–92.

Lin, T.-H., Tsao, Y., Wang, Y.-H., Yen, H.-W., & Lu, S.-S. (2017). Computing biodiversity change via a soundscape monitoring network. 2017 Pacific Neighborhood Consortium Annual Conference and Joint Meetings (PNC).

Lin, T.-H., Yu, H.-Y., Chen, C.-F., & Chou, L.-S. (2015). Passive Acoustic Monitoring of the Temporal Variability of Odontocete Tonal Sounds from a Long-Term Marine Observatory. *Plos One, 10*(4).

Lins, O. G., Picton, T. W., Boucher, B. L., Durieux-Smith, A., Champagne, S. C., Moran, L. M., Perez-Abalo, M. C., Martin, V., & Savio, G. (1996). Frequency-specific audiometry using steady-state responses. *Ear Hear. 17*, 81–96.

Listovsky, R. (2015). Development of the auditory system. *Handbook of clinical neurology, 129*, 55–72.

Llanos, F., Xie, Z., & Chandrasekaran, B. (2017). Hidden Markov modeling of frequency-following responses to Mandarin lexical tones. *Journal of Neuroscience Methods, 291*, 101–112.

Marsh, J. T., Brown, W. S., & Smith, J. C. (1974). Differential brainstem pathways for the conduction of auditory frequency-following responses. Electroencephalography and clinical neurophysiology, 36(4), 415–424.

Marsland, S. (2015). Machine Learning, 2nd Edition (2nd ed.). Retrieved from http://proquest.safaribooksonline.com.proxy.library.ohio.edu/9781466583283

Moushegian, G., Rupert, A. L., & Stillman, R. D. (1973). Scalp-recorded early responses in man to frequencies in the speech range. *Electroencephalography & Clinical Neurophysiology, 35*(6), 665–667.

Musacchia, G., Sams, M., Skoe, E., & Kraus, N. (2007). Musicians have enhanced subcortical auditory and audiovisual processing of speech and music. *Proceedings of the National Academy of Sciences of the United States of America, 104*, 15894–15898.

Oxenham, A. J. (2008). Pitch Perception and Auditory Stream Segregation: Implications for Hearing Loss and Cochlear Implants. *Trends in Amplification, 12*(4), 316-331.

Oxenham, A. J. (2012). Pitch perception. *The Journal of neuroscience: the official journal of the Society for Neuroscience, 32*(39), 13335–13338.

Parbery-Clark, A., Skoe, E., & Kraus, N. (2009). Musical experience limits the degradative effects of background noise on the neural processing of sound. *Journal of Neurosciences, 29*, 14100-14107.

Plyler, P. N., & Ananthanarayan, A. K. (2001). Human frequency-following responses: representation of second formant transitions in normal-hearing and hearing-impaired listeners. *Journal of the American Academy of Audiology, 12*(10), 523–533.

Ramus, F. (2014). Neuroimaging sheds new light on the phonological deficit in dyslexia. *Trends in Cognitive Sciences, 18*(6), 274–275.

Ramus, F., & Ahissar, M. (2012). Developmental dyslexia: The difficulties of interpreting poor performance, and the importance of normal performance. *Cognitive Neuropsychology, 29*(1–2), 104–122.

Ribas-Prats, T., Almeida, L., Costa-Faidella, J., Plana, M., Corral, M., Gómez-Roig, M. D., & Escera, C. (2019). The frequency-following response (FFR) to speech stimuli: A normative dataset in healthy newborns. *Hearing Research, 371*, 28–39.

Rickman, M. D., Chertoff, M. E., & Hecox, K. E. (1991). Electrophysiological evidence

of nonlinear distortion products to two-tone stimuli. *The Journal of the Acoustical

Society of America*, 89(6), 2818–2826.

Rocha-Muniz, C. N., Befi-Lopes, D. M., & Schochat, E. (2012). Investigation of auditory

processing disorder and language impairment using the speech-evoked auditory

brainstem response. *Hearing Research, 294*(1–2), 143–152.

Ruchkin, D. S., (1988). Measurement of event-related potentials: Signal extraction. In T.

W. Picton (Ed.), Handbook of electroencephalography and clinical neurophysiology

(Vol. 3, pp. 7–43). Amsterdam: Elsevier.

Russo, N. M., Skoe, E., Trommer, B., Nicol, T., Zecker, S., Bradlow, A., & Kraus, N.

(2008). Deficient brainstem encoding of pitch in children with Autism Spectrum

Disorders. *Clinical Neurophysiology: Official Journal of the International Federation

of Clinical Neurophysiology, 119*(8), 1720–1731.

Russo, N., Nicol, T., Trommer, B., Zecker, S., & Kraus, N. (2009). Brainstem

transcription of speech is disrupted in children with autism spectrum disorders.

Developmental Science, 12(4), 557–567.

Schimmel, H., Rapin, I., & Cohen, M. M. (1974). Improving evoked response audiometry

with special reference to the use of machine scoring. *International Journal of

Audiology, 13*(1), 33-65.

Schimmel, H., Rapin, I., & Cohen, M. M. (1975). Improving evoked response

audiometry. *International Journal of Audiology, 14*(5-6), 466-479.

Sininger, Yvonne. (1993). Auditory brainstem response for objective measures of

hearing. *Ear & Hearing, 14*(1), 23-30.

Skoe, E., & Kraus, N. (2010). Auditory brainstem response to complex sounds: A tutorial. *Ear and Hearing, 31*(3), 302–324.

Smalt, C. J., Krishnan, A., Bidelman, G. M., Ananthakrishnan, S., & Gandour, J. T. (2012). Distortion products and their influence on representation of pitch-relevant information in the human brainstem for unresolved harmonic complex tones. *Hearing research, 292*(1-2), 26–34.

Smith, J. C., Marsh, J. T., & Brown, W. S. (1975). Far-field recorded frequency-following responses: Evidence for the locus of brainstem sources. *Electroencephalography and Clinical Neurophysiology*, 39(5), 465–472.

Song, J. H., Banai, K., Russo, N. M., & Kraus, N. (2006). On the relationship between speech- and nonspeech-evoked auditory brainstem responses. *Audiology & Neuro-Otology, 11*(4), 233–241.

Song, J. H., Nicol, T., & Kraus, N. (2011). Test–retest reliability of the speech-evoked auditory brainstem response. *Clinical Neurophysiology, 122*(2), 346–355.

Tallal, P. (2004). Improving language and literacy is a matter of time. *Nature Reviews. Neuroscience, 5*(9), 721–728.

Valdes, J. L., Perez-Abalo, M. C., Martin, V., Savio, G., Sierra, C., Rodriguez, E., & Lins, O. (1997). Comparison of Statistical Indicators for the Automatic Detection of 80 Hz Auditory Steady State Responses. *Ear and Hearing, 18*(5), 420–429.

Van Dyke, K. B., Lieberman, R., Presacco, A., & Anderson, S. (2017). Development of Phase Locking and Frequency Representation in the Infant Frequency-Following Response. Journal of Speech, Language, and Hearing Research, 60(9), 2740–2751.

Vander Werff, K. R. (2009). Accuracy and Time Efficiency of Two ASSR Analysis Methods Using Clinical Test Protocols. *Journal of the American Academy of Audiology, 20*(7), 433-452.

Wall, C., Simard, P., Lembke, C., & Mann, D. (2013). Large-scale passive acoustic monitoring of fish sound production on the West Florida Shelf. *Marine Ecology Progress Series, 484*, 173–188.

Walters, C. L., Collen, A., Lucas, T., Mroz, K., Sayer, C. A., & Jones, K. E. (2013). Challenges of Using Bioacoustics to Globally Monitor Bats. *Bat Evolution, Ecology, and Conservation*, 479–499.

Wimmer, J.D., (2015) "Acoustic sensing: roles and applications in monitoring avian biodiversity," Ph.D. book, Queensland University of Technology.

Wong, P. C. M., Skoe, E., Russo, N. M., Dees, T., & Kraus, N. (2007). Musical experience shapes human brainstem encoding of linguistic pitch patterns. *Nature Neuroscience, 10*, 420–422.

Wong, P. K., & Bickford, R. G. (1980). Brain stem auditory evoked potentials: The use of noise estimate. *Electroencephalography and Clinical Neurophysiology, 50*, 25- 34.

Worden, F. G., & Marsh, J. T. (1968). Frequency-following (microphonic-like) neural responses evoked by sound. Electroencephalography and clinical neurophysiology, 25(1), 42–52.

Xie, Z., Reetzke, R., &; Chandrasekaran, B. (2019). Machine learning approaches to analyze speech-evoked neurophysiological responses. Journal of Speech, Language, and Hearing Research, 62(3), 587–601.

Yu, X., Hu, D., & Xu, J. (2014). Non-negative Matrix Factorization Algorithms and Applications. In Blind source separation: theory and applications (pp. 245–311). essay, John Wiley & Sons Singapore Pte. Ltd.

Zhang, X., & Gong, Q. (2019). Frequency-Following Responses to Complex Tones at Different Frequencies Reflect Different Source Configurations. *Frontiers in Neuroscience, 13*, 130.

Appendix A

Adult NMF Separation

Graphical presentations of the NMF separation procedure for an adult participant at 5 iterations.

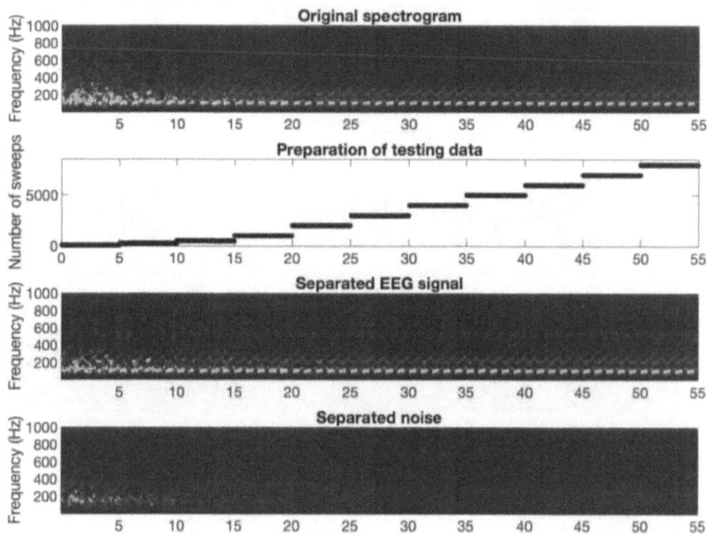

Graphical presentations of the NMF separation procedure for an adult participant at 10 iterations.

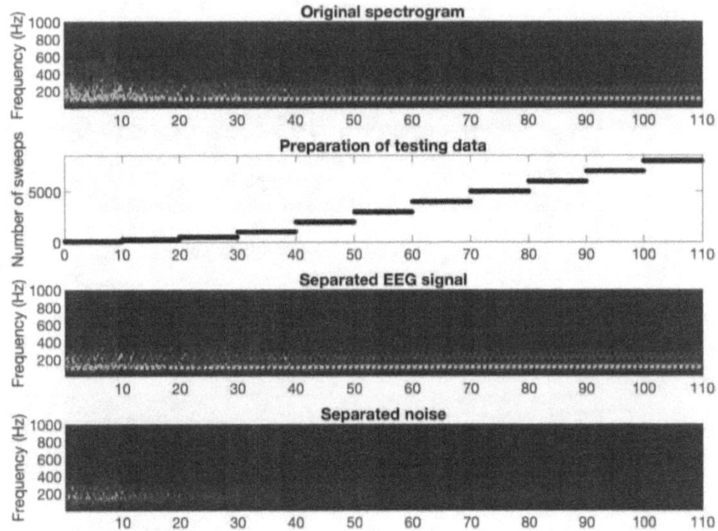

Graphical presentations of the NMF separation procedure for an adult participant at 100 iterations.

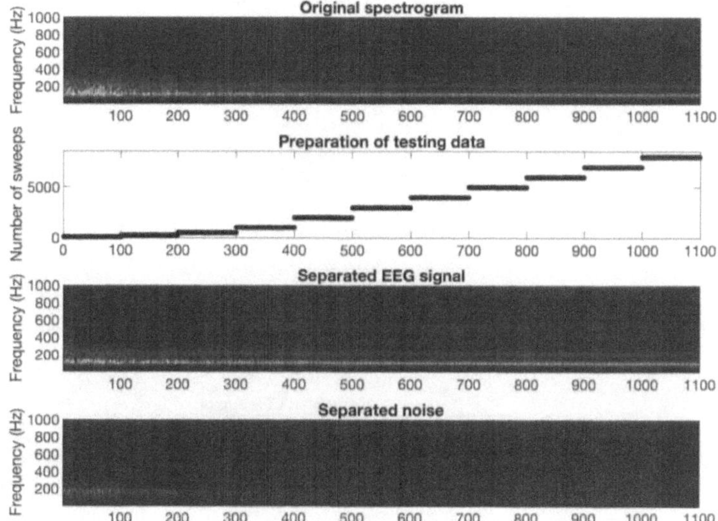

Appendix B

Newborn NMF Separation

Graphical presentations of the NMF separation procedure for a newborn participant at 5 iterations.

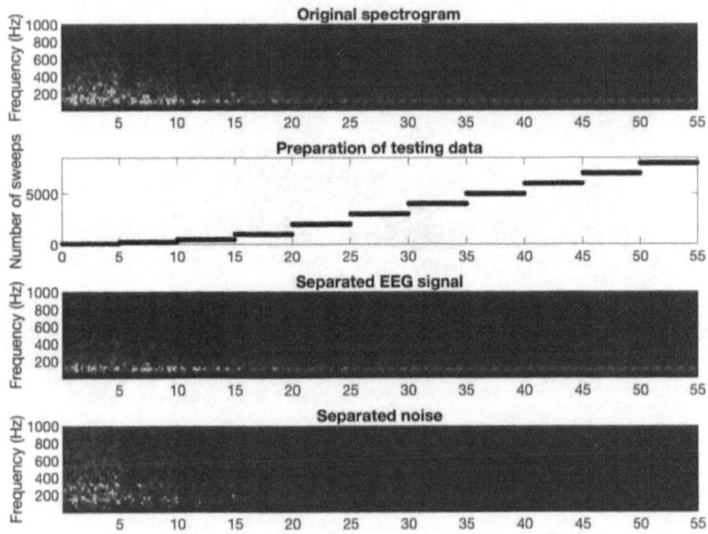

Graphical presentations of the NMF separation procedure for a newborn participant at 10 iterations.

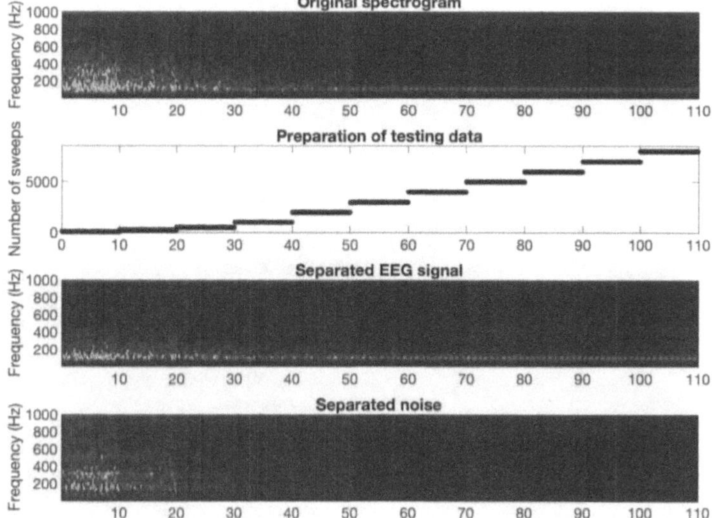

Graphical presentations of the NMF separation procedure for a newborn participant at 100 iterations

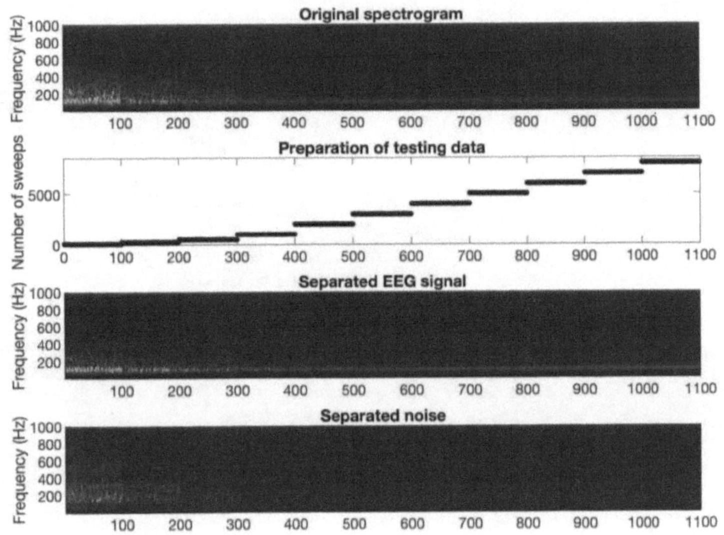

Appendix C

1 Iteration nSweeps Confusion Matrices

Pairwise comparison table depicting results based on the results of a one-way ANOVA depicting the relationship between Frequency Error and the nSweeps condition for the hidden component of newborn and adult data at one iteration.

FE	1	2	3	4	5	6	7	8	9	10	11
1		-.231 (± .774)	.966 (± . 763)	2.136* (± .576)	3.709** (± .659)	3.297** (± .686)	4.263** (± .663)	4.666** (± .674)	4.895** (± .622)	5.075** (± .625)	5.054** (± .627)
2	.231 (± .774)		1.197 (± .757)	2.367 (± .678)	3.940** (± .635)	3.528** (± .734)	4.494** (± .579)	4.897** (± .679)	5.126** (± .595)	5.306** (± .620)	5.285** (± .624)
3	-.966 (± .763)	-1.197 (± .757)		1.170 (± . 497)	2.743** (± .594)	2.331 (± .697)	3.297** (± .594)	3.700** (± .550)	3.929** (± .552)	4.109** (± .550)	4.088** (± .545)
4	-2.136* (± .576)	-2.367 (± .678)	-1.170 (± .497)		1.572 (± .468)	1.160 (± .484)	2.127** (± .413)	2.529** (± .420)	2.758** (± .383)	2.938** (± .369)	2.918** (± .370)
5	-3.709** (± .659)	-3.940** (± .635)	-2.743** (± .594)	-1.572 (± .468)		-.412 (± .486)	.555 (± .302)	.957 (± .372)	1.186* (± .332)	1.366** (± .334)	1.345* (± .344)
6	-3.297** (± .686)	-3.528** (± .734)	-2.331 (± .697)	-1.160 (± .484)	.412 (± .468)		.967 (± .461)	1.369 (± .425)	1.598** (± .394)	1.778** (± .410)	1.757** (± .401)
7	-4.263** (± .663)	-4.494** (± .579)	-3.297** (± .594)	-2.127** (± .413)	-.555 (± .302)	-.967 (± .461)		.402 (± .324)	.631 (± .247)	.811* (± .221)	.790 (± .228)
8	-4.666** (± .674)	-4.897** (± .679)	-3.700** (± .550)	-2.529** (± .420)	-.957 (± .372)	-1.369 (± .425)	-.402 (± .324)		.229 (± .307)	.409 (± .200)	.388 (± .212)
9	-4.895** (± .622)	-5.126** (± .595)	-3.929** (± .552)	-2.758** (± .383)	-1.186* (± .332)	-1.598** (± .394)	-.631 (± .247)	-.229 (± .307)		.180 (± .180)	.159 (± .184)
10	-5.075** (± .625)	-5.306** (± .620)	-4.109** (± .550)	-2.938** (± .369)	-1.366** (± .334)	-1.778** (± .410)	-.811* (± .221)	-.409 (± .200)	-.180 (± .180)		-.021 (± .108)
11	-5.054** (± .627)	-5.285** (± .624)	-4.088** (± .545)	-2.918** (± .370)	-1.345* (± .344)	-1.757** (± .401)	-.790 (± .228)	-.388 (± .212)	-.159 (± .184)	.021 (± .108)	

**P<.01 * P<.05

Pairwise comparison table depicting results based on the results of a one-way ANOVA depicting the relationship between Slope Error and the nSweeps condition for the hidden component of newborn and adult data at one iteration.

SE	1	2	3	4	5	6	7	8	9	10	11
1		-11.973 (±19.205)	-49.758 (± 15.297)	- 77.637** (±18.399)	- 71.840** (±16.811)	- 79.953** (±16.791)	- 103.147** (±15.978)	- 100.471** (±16.915)	- 110.321** (±14.361)	- 112.566** (±15.328)	- 110.252** (±15.602)
2	11.973 (±19.205)		-37.785 (±17.230)	- 65.663** (±13.751)	-59.867 (± 18.632)	- 67.979** (±15.720)	-91.174** (±14.054)	-88.498** (±15.554)	-98.347** (±14.096)	- 100.593** (±14.741)	-98.279** (±14.342)
3	49.758** (± 15.297)	37.785 (±17.230)		-27.879 (±13.069)	-22.082 (± 12.152)	-30.195 (±13.353)	-53.389** (±11.815)	-50.713* (±13.619)	-60.563** (±11.604)	-62.808** (±11.403)	-60.494** (±11.597)
4	77.637** (± 18.399)	65.663** (±13.751)	27.879 (±13.069)		5.796 (±14.890)	-2.316 (±11.006)	-25.511 (±9.623)	-22.834 (±9.355)	-32.684 (±9.503)	-34.929** (±8.663)	-32.615** (±8.091)
5	71.840** (± 16.811)	59.867 (± 18.632)	22.082 (±12.152)	-5.796 (±14.890)		-8.112 (±12.621)	-31.307 (±10.924)	-28.631 (±13.263)	-38.480 (±11.685)	-40.726 (±11.871)	-38.412 (±11.944)
6	79.953** (± 16.791)	67.979** (±15.720)	30.195 (±13.353)	2.316 (±11.006)	8.112 (±12.621)		-23.195 (±10.874)	-20.518 (±9.206)	-30.368* (±8.563)	-32.613* (±8.683)	-30.299 (±9.014)
7	103.147** (± 15.978)	91.174** (±14.054)	53.389** (±11.815)	25.511 (±9.623)	31.307 (±10.924)	23.195 (±10.874)		2.676 (±8.646)	-7.173 (± 5.756)	-9.419 (±6.643)	-7.105 (±6.354)
8	100.471** (±16.915)	88.498** (±15.554)	50.713* (±13.619)	22.834 (±9.355)	28.631 (±13.263)	20.518 (±9.206)	-2.676 (±8.646)		-9.849 (±6.930)	-12.095 (±5.104)	-9.781 (±4.399)
9	110.321** (± 14.361)	98.347** (±14.096)	60.563** (±11.604)	32.684 (±9.503)	38.480 (±11.685)	30.368* (±8.563)	7.173 (±5.756)	9.849 (± 6.930)		-2.245 (±5.498)	.069 (±4.565)
10	112.566** (±15.328)	100.593** (± 14.741)	62.808** (±11.403)	34.929** (±8.663)	40.726 (±11.871)	32.613* (±8.683)	9.419 (±6.643)	12.095 (± 5.104)	2.245 (±5.498)		2.314 (±3.125)
11	110.252** (±15.602)	98.279** (± 14.342)	60.494 (±11.597)	32.615** (±8.091)	38.412 (±11.944)	30.299 (±9.014)	7.105 (±6.354)	9.781 (±4.399)	-.069 (±4.565)	-2.314 (±3.125)	

**P<.01 * P<.05

Pairwise comparison table depicting results based on the results of a one-way ANOVA depicting the relationship between Tracking Accuracy and the nSweeps condition for the hidden component of newborn and adult data at one iteration.

TA	1	2	3	4	5	6	7	8	9	10	11
1		.022 (±.049)	-.028 (±.051)	-.094 (±.045)	-.151 (±.044)	-.100 (±.049)	-.197** (±.039)	-.201** (±.045)	-.227** (±.036)	-.231** (±.040)	-.233** (±.040
2	-.022 (± .049)		-.050 (±.044)	-.116 (±.041)	-.173* (± .044)	-.122 (±.053)	-.219** (± .040)	-.223** (±.045)	-.249** (±.038)	-.253** (±.041)	-.255** (±.040)
3	.028 (±.051)	.050 (±.044)		-.066 (±.039)	-.123 (± .039)	-.072 (±.047)	-.169** (±.037)	-.173** (±.040)	-.199** (±.034)	-.202** (±.038)	-.205** (±.037)
4	.094 (±.045)	.116 (±.041)	.066 (±.039)		-.057 (±.034)	-.006 (±.036)	-.103 (±.030)	-.107* (±.030)	-.133** (±.027)	-.137** (±.027)	-.139** (±.027)
5	.151 (±.044)	.173* (± .044)	.123 (±.039)	.057 (±.034)		.051 (±.036)	-.046 (± .029)	-.050 (±.032)	-.076 (±.030)	-.080 (±.030)	-.082 (±.030)
6	.100 (±.049)	.122 (±.053)	.072 (±.047)	.006 (±.036)	-.051 (±.036)		-.097 (±.035)	-.101 (±.029)	-.127 (±.031)	-.131** (±.029)	-.133** (±.029)
7	.197** (±.039)	.219** (±.040)	.169** (±.037)	.103 (±.030)	.046 (±.029)	.097 (± .035)		-.004 (±.030)	-.030 (±.017)	-.034 (±.022)	-.036 (±.022)
8	.201** (±.045)	.223** (±.045)	.173** (±.040)	.107 (±.030)	.050 (±.032)	.101 (±.029)	.004 (±.030)		-.026 (±.023)	-.029 (±.016)	-.032 (±.016)
9	.227** (±.036)	.249** (±.038)	.199** (± .034)	.133** (±.027)	.076 (±.030)	.127** (± .031)	.030 (±.017)	.026 (±.023)		-.004 (±.015)	-.006 (±.013)
10	.231** (±.040)	.253** (±.041)	.202** (±.038)	.137** (±.027)	.080 (±.030)	.131** (±.029)	.034 (±.022)	.029 (±.016)	.004 (±.015)		-.002 (±.009)
11	.233** (±.040)	.255=** (±.040)	.205** (±.037)	.139** (±.027)	.082 (±.030)	.133** (±.029)	.036 (±.022)	.032 (±.016)	.006 (±.013)	.002 (±.009)	

**P<.01 * P<.05

Pairwise comparison table depicting results based on the results of a one-way ANOVA depicting the relationship between Frequency Error and the nSweeps condition for the hidden component of adult data at one iteration.

FE	1	2	3	4	5	6	7	8	9	10	11
1		-.196 (±1.045)	2.411 (±1.161)	2.690 (±.881)	4.274** (±.800)	4.113** (±.816)	4.963** (±.817)	4.930** (±.861)	4.860** (±.781)	5.182** (±.770)	5.061** (±.789)
2	.196 (±1.045)		2.607 (±.961)	2.886 (±.979)	4.470** (±.862)	4.309* (±1.065)	5.159** (±.866)	5.126** (±.944)	5.056** (±.926)	5.377** (±.922)	5.257** (±.905)
3	-2.411 (±1.161)	-2.607 (±.961)		.279 (±.648)	1.863 (±.816)	1.702 (±.910)	2.552 (±.710)	2.519 (±.729)	2.449 (±.749)	2.771* (±.738)	2.650 (±.722)
4	-2.690 (±.881)	-2.886 (±.979)	-.279 (±.648)		1.584 (±.617)	1.423 (±.536)	2.273** (±.436)	2.240** (±.492)	2.170** (±.437)	2.492** (±.459)	2.371** (±.432)
5	-4.274** (±.800)	-4.470** (±.862)	-1.863 (±.816)	-1.584 (±.617)		-.161 (±.661)	.689 (±.401)	.656 (±.525)	.586 (±.452)	.908 (±.459)	.787 (±.497)
6	-4.113** (±.816)	-4.309* (±1.065)	-1.702 (±.910)	-1.423 (±.536)	.161 (±.661)		.850 (±.523)	.817 (±.507)	.746 (±.345)	1.068 (±.436)	.948 (±.384)
7	-4.963** (±.817)	-5.159** (±.866)	-2.552 (±.710)	-2.273** (±.436)	-.689 (±.401)	-.850 (±.523)		-.034 (±.334)	-.104 (±.264)	.218 (±.229)	.098 (±.241)
8	-4.930** (±.861)	-5.126** (±.944)	-2.519 (±.729)	-2.240** (±.492)	-.656 (±.525)	-.817 (±.507)	.034 (±.334)		-.070 (±.269)	.252 (±.228)	.131 (±.262)
9	-4.860** (±.781)	-5.056** (±.926)	-2.449 (±.749)	-2.170** (±.437)	-.586 (±.452)	-.746 (±.345)	.104 (±.264)	.070 (±.269)		.322 (±.197)	.201 (±.134)
10	-5.182** (±.770)	-5.377** (±.922)	-2.771* (±.738)	-2.492** (±.459)	-.908 (±.459)	-1.068 (±.436)	-.218 (±.229)	-.252 (±.228)	-.322 (±.197)		-.121 (±.160)
11	-5.061** (±.789)	-5.257** (±.905)	-2.650 (±.722)	-2.371** (±.432)	-.787 (±.497)	-.948 (±.384)	-.098 (±.241)	-.131 (±.262)	-.201 (±.134)	.121 (±.160)	

**P<.01 * P<.05

Pairwise comparison table depicting results based on the results of a one-way ANOVA depicting the relationship between Slope Error and the nSweeps condition for the hidden component of adult data at one iteration.

SE	1	2	3	4	5	6	7	8	9	10	11
1		-2.425 (±23.34)	-61.457 (±19.36)	-70.407 (±21.233)	-70.407 (±22.123)	-83.068* (±21.637)	-93.44** (±19.261)	-95.880** (±18.964)	-99.894** (±18.563)	-102.42 ** (±19.160)	-98.33** (±18.848)
2	2.425 (±23.346)		-59.032 (±17.490)	-67.983* (±17.955)	-67.983 (±20.298)	-80.643* (±20.040)	-91.024** (±17.230)	-93.455** (±20.661)	-97.469** (±19.038)	-99.997** (±19.186)	-95.899** (±17.904)
3	61.457 (±19.369)	59.032 (±17.490)		-8.951 (±14.885)	-8.951 (±12.830)	-21.611 (±12.077)	-31.992 (±13.111)	-34.423 (±11.628)	-38.437 (±11.690)	-40.965 (±11.384)	-36.866 (±11.515)
4	75.244 (±21.233)	72.820* (±17.955)	13.788 (±14.885)		4.837 (±15.555)	-7.823 (±12.660)	-18.204 (±9.073)	-20.636 (±11.435)	-24.649 (±10.123)	-27.178 (±9.764)	-23.079 (±9.232)
5	70.407 (±22.123)	67.983 (±20.298)	8.951 (±12.830)	4.837 (±15.555)		-12.660 (±15.520	-23.041 (±13.029)	-25.473 (±12.704)	-29.486 (±13.337)	-32.015 (±13.144)	-27.916 (±13.167)
6	83.068* (±21.637)	80.643* (±20.040)	21.611 (±)12.077	12.660 (±12.660)	12.660 (±15.520)		-10.381 (±14.477)	-12.812 (±10.311)	-16.826 (±10.369)	-19.354 (±11.006)	-15.255 (±10.666)
7	93.449** (±19.261)	91.024** (±17.230)	31.992 (±13.111)	23.041 (±9.073)	23.041 (±13.029)	10.381 (±14.477)		-2.431 (±8.930)	-6.445 (±7.370)	-8.973 (±6.512)	-4.874 (±6.160)
8	95.880** (±18.964)	93.455** (±20.661)	34.423 (±11.628)	25.473 (±11.435)	25.473 (±12.704)	12.812 (±10.311)	2.431 (±8.930)		-4.014 (±4.401)	-6.542 (±5.054)	-2.443 (±5.202)
9	99.894** (±18.563)	97.469** (±19.038)	38.437 (±11.690)	29.486 (±)10.123	29.486 (±13.337)	16.826 (±10.369)	6.445 (±7.370)	4.014 (±4.401)		-2.528 (±4.550)	1.571 (±2.600)
10	102.422** (±19.160)	99.997** (±19.186)	40.965 (±11.384)	32.015 (±9.764)	32.015 (±13.144)	19.354 (±11.006)	8.973 (±6.512)	6.542 (±5.054)	2.528 (±4.550)		4.099 (±3.734)
11	98.323** (±18.848)	95.899** (±17.904)	36.866 (±11.515)	27.916 (±9.232)	27.916 (±13.167)	15.255 (±10.666)	4.874 (±6.160)	2.443 (±5.202)	-1.571 (±2.600)	-4.099 (±3.734)	

**P<.01 * P<.05

Pairwise comparison table depicting results based on the results of a one-way ANOVA depicting the relationship between Tracking Accuracy and the nSweeps condition for the hidden component of newborn and adult data at one iteration.

TA	1	2	3	4	5	6	7	8	9	10	11
1		-.031 ($\pm$.072)	-.156 ($\pm$.057)	-.196 ($\pm$.057)	-.239** ($\pm$.046)	-.232* ($\pm$.061)	-.289** ($\pm$.046)	-.287** ($\pm$.054)	-.299** ($\pm$.050)	-.311** ($\pm$.050)	-.309** ($\pm$.052)
2	.031 ($\pm$.072)		-.125 ($\pm$.053)	-.165 ($\pm$.048)	-.208* ($\pm$.052)	-.201 ($\pm$.065)	-.258** ($\pm$.056)	-.256** ($\pm$.055)	-.268** ($\pm$.053)	-.280** ($\pm$.054)	-.278** ($\pm$.049)
3	.156 ($\pm$.057)	.125 ($\pm$.053)		-.040 ($\pm$.030)	-.082 ($\pm$.040)	-.075 ($\pm$.038)	-.133 ($\pm$.039)	-.131* ($\pm$.034)	-.143* ($\pm$.035)	-.155** ($\pm$.035)	-.153** ($\pm$.032)
4	.196 ($\pm$.057)	.165 ($\pm$.048)	.040 ($\pm$.030)		-.043 ($\pm$.039)	-.036 ($\pm$.034)	-.093 ($\pm$.028)	-.091 ($\pm$.032)	-.103* ($\pm$.027)	-.115* ($\pm$.027)	-.113** ($\pm$.025)
5	.239** ($\pm$.046)	.208* ($\pm$.052)	.082 ($\pm$.040)	.043 ($\pm$.039)		.007 ($\pm$.047)	-.050 ($\pm$.034)	-.048 ($\pm$.038)	-.061 ($\pm$.035)	-.072 ($\pm$.033)	-.070 ($\pm$.034)
6	.232* ($\pm$.061)	.201 ($\pm$.065)	.075 ($\pm$.038)	.036 ($\pm$.034)	-.007 ($\pm$.047)		-.057 ($\pm$.033)	-.055 ($\pm$.034)	-.068 ($\pm$.029)	-.079 ($\pm$.026)	-.077 ($\pm$.029)
7	.289** ($\pm$.046)	.258** ($\pm$.056)	.133 ($\pm$.039)	.093 ($\pm$.028)	.050 ($\pm$.034)	.057 ($\pm$.033)		.002 ($\pm$.029)	-.010 ($\pm$.017)	-.022 ($\pm$.021)	-.020 ($\pm$.021)
8	.287** ($\pm$.054)	.256** ($\pm$.055)	.131* ($\pm$.034)	.091 ($\pm$.032)	.048 ($\pm$.038)	.055 ($\pm$.034)	-.002 ($\pm$.029)		-.012 ($\pm$.022)	-.024 ($\pm$.019)	-.022 ($\pm$.020)
9	.299** ($\pm$.050)	.268** ($\pm$.053)	.143* ($\pm$.035)	.103* ($\pm$.027)	.061 ($\pm$.035)	.068 ($\pm$.029)	.010 ($\pm$.017)	.012 ($\pm$.022)		-.012 ($\pm$.017)	-.010 ($\pm$.010)
10	.311** ($\pm$.050)	.280** ($\pm$.054)	.155** ($\pm$.035)	.115* ($\pm$.027)	.072 ($\pm$.033)	.079 ($\pm$.026)	.022 ($\pm$.021)	.024 ($\pm$.019)	.012 ($\pm$.017)		.002 ($\pm$.014)
11	.309** ($\pm$.052)	.278** ($\pm$.049)	.153** ($\pm$.032)	.113** ($\pm$.025)	.070 ($\pm$.034)	.077 ($\pm$.029)	.020 ($\pm$.021)	.022 ($\pm$.020)	.010 ($\pm$.010)	-.002 ($\pm$.014)	

**P<.01 * P<.05

Pairwise comparison table depicting results based on the results of a one-way ANOVA depicting the relationship between Frequency Error and the nSweeps condition for the hidden component of newborn data at one iteration.

FE	1	2	3	4	5	6	7	8	9	10	11
1		-.266 (±1.128)	-.479 (±1.002)	1.583 (±.754)	3.143 (±1.022)	2.480 (±1.074)	3.564 (±1.020)	4.402** (±1.018)	4.929** (±.947)	4.968** (±.962)	5.047** (±.954)
2	.266 (±1.128)		-.213 (±1.148)	1.849 (±.936)	3.409* (±.921)	2.746 (±1.011)	3.830** (±.775)	4.668** (±.969)	5.196** (±.760)	5.234** (±.835)	5.313** (±.859)
3	.479 (±1.002)	.213 (±1.148)		2.062 (±.741)	3.622* (±.854)	2.959 (±1.038)	4.043** (±.928)	4.881** (±.811)	5.409** (±.801)	5.447** (±.805)	5.526** (±.805)
4	-1.583 (±.754)	-1.849 (±.936)	-2.062 (±.741)		1.560 (±.693)	.897 (±.781)	1.981 (±.679)	2.819* (±.662)	3.347* (±.611)	3.385** (±.565)	3.464** (±.583)
5	-3.143 (±1.022)	-3.409 (±.921)	-3.622* (±.854)	-1.560 (±.693)		-.663 (±.705)	.420 (±.446)	1.258 (±.525)	1.786 (±.480)	1.825* (±.481)	1.904* (±.476)
6	-2.480 (±1.074)	-2.746 (±1.011)	-2.959 (±1.038)	-.897 (±.781)	.663 (±.705)		1.083 (±.738)	1.921 (±.666)	2.449 (±.680)	2.488* (±.671)	2.567* (±.679)
7	-3.564 (±1.020)	-3.830** (±.775)	-4.043** (±.928)	-1.981 (±.679)	-.420 (±.446)	-1.083 (±.738)		.838 (±.535)	1.366 (±.404)	1.404* (±.366)	1.483* (±.375)
8	-4.402** (±1.018)	-4.668** (±.969)	-4.881** (±.811)	-2.819* (±.662)	-1.258 (±.525)	-1.921 (±.666)	-.838 (±.535)		.528 (±.530)	.566 (±.320)	.645 (±.326)
9	-4.929** (±.947)	-5.196** (±.760)	-5.409** (±.801)	-3.347** (±.611)	-1.786* (±.480)	-2.449 (±.680)	-1.366 (±.404)	-.528 (±.530)		.038 (±.293)	.117 (±.326)
10	-4.968** (±.962)	-5.234** (±.835)	-5.447** (±.805)	-3.385** (±.565)	-1.825* (±.481)	-2.488* (±.671)	-1.404* (±.366)	-.566 (±.320)	-.038 (±.293)		.079 (±.145)
11	-5.047** (±.954)	-5.313** (±.859)	-5.526** (±.805)	-3.464** (±.583)	-1.904* (±.476)	-2.567* (±.679)	-1.483* (±.375)	-.645 (±.326)	-.117 (±.326)	-.079 (±.145)	

**P<.01 * P<.05

Pairwise comparison table depicting results based on the results of a one-way ANOVA depicting the relationship between Slope Error and the nSweeps condition for the hidden component of newborn data at one iteration.

SE	1	2	3	4	5	6	7	8	9	10	11
1		-21.522 (±29.758)	-38.060 (±23.198)	-80.029 (±29.201)	-73.273 (±24.913)	-76.838 (±25.207)	- 112.846** (±24.861)	- 105.062* (±27.165)	- 120.748** (±21.519)	- 122.710** (±23.410)	- 122.181** (±24.251)
2	21.522 (±29.758)		-16.537 (±28.635)	-58.507 (±20.479)	-51.751 (±30.249)	-55.316 (±23.749)	-91.324* (±21.684)	-83.540 (±22.911)	-99.225** (±20.532)	- 101.188** (±21.997)	- 100.659** (±21.919)
3	38.060 (±23.198)	16.537 (±28.635)		-41.970 (±20.856)	-35.214 (±19.947)	-38.778 (±22.870)	-74.787* (±19.051)	-67.003 (±23.609)	-82.688** (±19.329)	-84.651** (±19.045)	-84.121** (±19.398)
4	80.029 (±29.201)	58.507 (±20.479)	41.970 (±20.856)		6.756 (±24.525)	3.191 (±17.491)	-32.817 (±16.320)	-25.033 (±14.456)	-40.718 (±15.551)	-42.681 (±13.883)	-42.152 (±12.902)
5	73.273 (±24.913)	51.751 (±30.249)	35.214 (±19.947)	-6.756 (±24.525)		-3.565 (±19.443)	-39.573 (±17.083)	-31.789 (±22.403)	-47.474 (±18.632)	-49.437 (±19.157)	-48.907 (±19.306)
6	76.838 (±25.207)	55.316 (±23.749)	38.778 (±22.870)	-3.191 (±17.491)	3.565 (±19.443)		-36.008 (±15.993)	-28.224 (±14.791)	-43.909 (±13.295)	-45.872 (±13.160)	-45.343 (±14.148)
7	112.846* (±24.861)	91.324* (±21.684)	74.787 (±19.051)	32.817 (±16.320)	39.573 (±17.083)	36.008 (±15.993)		7.784 (±14.293)	-7.901 (±8.672)	-9.864 (±11.152)	-9.335 (±10.700)
8	105.062 (±27.165)	83.540 (±22.911)	67.003 (±23.609)	25.033 (±14.456)	31.789 (±22.403)	28.224 (±14.791)	-7.784 (±14.293)		-15.685 (±12.528)	-17.648 (±8.544)	-17.119 (±6.907)
9	122.181** (±21.519)	99.225** (±20.532)	82.688 (±19.329)	40.718 (±15.551)	47.474 (±18.632)	43.909 (±13.295)	7.901 (±8.672)	15.685 (±12.528)		-1.963 (±9.588)	-1.433 (±8.333)
10	120.748** (±23.410)	101.188** (±21.997)	84.651 (±19.045)	42.681 (±13.883)	49.437 (±19.157)	45.872 (±13.160)	9.864 (±11.152)	17.648 (±8.544)	1.963 (±9.588)		.529 (±4.884)
11	122.710** (±24.251)	100.659** (±21.919)	84.121 (±19.398)	42.152 (±12.902)	48.907 (±19.306)	45.343 (±14.148)	9.335 (±10.700)	17.119 (±6.907)	1.433 (±8.333)	-.529 (±4.884)	

**P<.01 * P<.05

Pairwise comparison table depicting results based on the results of a one-way ANOVA depicting the relationship between Tracking Accuracy and the nSweeps condition for the hidden component of newborn and adult data at one iteration.

TA	1	2	3	4	5	6	7	8	9	10	11
1		.075 (±.067)	.099 (±.082)	.008 (±.067)	-.063 (±.073)	.032 (±.074)	-.105 (±.062)	-.116 (±.070)	-.155 (±.052)	-.151 (±.060)	-.157 (±.060)
2	-.075 (±.067)		.025 (±.069)	-.067 (±.064)	-.138 (±.069)	-.043 (±.081)	-.180 (±.058)	-.191 (±.070)	-.230* (±.054)	-.226 (±.062)	-.232* (±.061)
3	-.099 (±.082)	-.025 (±.069)		-.092 (±.068)	-.163 (±.065)	-.068 (±.083)	-.205 (±.061)	-.215 (±.069)	-.255** (±.057)	-.250* (±.064)	-.256* (±.063)
4	-.008 (±.067)	.067 (±.064)	.092 (±.068)		-.071 (±.054)	.024 (±.060)	-.113 (±.050)	-.124 (±.048)	-.163 (±.046)	-.158 (±.045)	-.165 (±.045)
5	.063 (±.073)	.138 (±.069)	.163 (±.065)	.071 (±.054)		.095 (±.052)	-.042 (±.045)	-.053 (±.050)	-.092 (±.048)	-.087 (±.049)	-.094 (±.048)
6	-.032 (±.074)	.043 (±.081)	.068 (±.083)	-.024 (±.060)	-.095 (±.052)		-.137 (±.060)	-.148 (±.046)	-.187 (±.053)	-.183 (±.049)	-.189* (±.049)
7	.105 (±.062)	.180 (±.058)	.205 (±.061)	.113 (±.050)	.042 (±.045)	.137 (±.060)		-.011 (±.050)	-.050 (±.029)	-.046 (±.038)	-.052 (±.038)
8	.116 (±.070)	.191 (±.070)	.215 (±.069)	.124 (±.048)	.053 (±.050)	.148 (±.046)	.011 (±.050)		-.039 (±.039)	-.035 (±.025)	-.041 (±.024)
9	.155 (±.052)	.230* (±.054)	.255** (±.057)	.163 (±.046)	.092 (±.048)	.187 (±.053)	.050 (±.029)	.039 (±.039)		.005 (±.024)	-.002 (±.023)
10	.151 (±.060)	.226 (±.062)	.250* (±.064)	.158 (±.045)	.087 (±.049)	.183 (±.049)	.046 (±.038)	.035 (±.025)	-.005 (±.024)		-.006 (±.011)
11	.157 (±.060)	.232* (±.061)	.256* (±.063)	.165 (±.045)	.094 (±.048)	.189* (±.049)	.052 (±.038)	.041 (±.024)	.002 (±.023)	.006 (±.011)	

**P<.01 * P<.05

Pairwise comparison table depicting results based on the results of a two-way ANOVA depicting the relationship between Frequency Error and the nSweeps condition for the newborn and adult data at one iteration.

FE	1	2	3	4	5	6	7	8	9	10	11
1		-1.924* (±.483)	-2.618** (±.526)	-3.861** (±.486)	-5.440** (±.440)	-5.974** (±).479	-6.650** (±.421)	-6.903** (±.459)	-6.862** (±.464)	-7.219** (±.456)	-7.268** (±.444)
2	1.924** (±.483)		-.694 (±.451)	-1.937* (±.502)	-3.516** (±.438)	-4.050** (±.499)	-4.726** (±.451)	-4.979** (±.452)	-4.938** (±.460)	-5.295** (±.469)	-5.344** (±.481)
3	2.618** (±.526)	.694* (±.451)		-1.243 (±.395)	-2.822** (±.485)	-3.356** (±.501)	-4.032** (±.483)	-4.285** (±.484)	-4.244** (±.458)	-4.601** (±.483)	-4.650** (±.512)
4	3.861** (±.486)	1.937** (±.502)	1.243 (±.395)		-1.579** (±.361)	-2.113** (±.394)	-2.789** (±.425)	-3.042** (±.460)	-3.001** (±.426)	-3.358** (±.447)	-3.407** (±.452)
5	5.440** (±.440)	3.516** (±.438)	2.822** (±.485)	1.579** (±.361)		-.534 (±.314)	-1.210 (±.348)	-1.463** (±.309)	-1.422** (±.326)	-1.780** (±.345)	-1.829** (±.330)
6	5.974** (±.479)	4.050** (±.499)	3.356** (±.501)	2.113** (±.394)	.534 (±.314)		-.676 (±.279)	-.929 (±.307)	-.888 (±.295)	-1.245** (±.302)	-1.294** (±.291)
7	6.650** (±.421)	4.726** (±.451)	4.032** (±.483)	2.789** (±.425)	1.210 (±.348)	.676 (±.279)		-.253 (±.265)	-.212 (±.261)	-.570 (±.253)	-.619 (±.253)
8	6.903** (±.459)	4.979** (±.452)	4.285** (±.484)	3.042** (±.460)	1.463** (±.309)	.929 (±.307)	.253 (±.265)		.041 (±.159)	-.317 (±.157)	-.366 (±.164)
9	6.862** (±.464)	4.938** (±.460)	4.244** (±.458)	3.001** (±.426)	1.422** (±.326)	.888 (±.295)	.212 (±.261)	-.041 (±.159)		-.358 (±.124)	-.407 (±.184)
10	7.219** (±.456)	5.295** (±.469)	4.601** (±.483)	3.358** (±.447)	1.780** (±.345)	1.245** (±.302)	.570 (±.253)	.317 (±.157)	.358 (±.124)		-.049 (±.118)
11	7.268** (±.444)	5.344** (±.481)	4.650** (±.512)	3.407** (±.452)	1.829** (±.330)	1.294** (±.291)	.619 (±.253)	.366 (±.164)	.407 (±.184)	.049 (±.118)	

**P<.01 * P<.05

214

Pairwise comparison table depicting results based on the results of a two-way ANOVA depicting the relationship between Slope Error and the nSweeps condition for the newborn and adult data at one iteration.

SE	1	2	3	4	5	6	7	8	9	10	11
1		39.039 (±13.894)	67.713** (±14.607)	97.757** (±12.987)	94.451** (±11.662)	104.672** (±11.274)	120.875** (±12.091)	121.426** (±11.874)	125.107** (±11.768)	129.845** (±11.406)	126.442** (±11.582)
2	-39.039 (±13.894)		28.675 (±11.514)	58.718** (±11.756)	55.413* (±14.037)	65.634** (±11.902)	81.836** (±11.853)	82.388** (±12.499)	86.069** (±11.970)	90.806** (±11.497)	87.403** (±11.302)
3	-67.713** (±14.607)	-28.675 (±11.514)		30.043 (±10.460)	26.738 (±11.853)	36.959 (±12.027)	53.161** (±10.511)	53.713* (±13.767)	57.394** (±11.881)	62.132** (±11.315)	58.728** (±11.383)
4	-97.757** (±12.987)	-58.718** (±11.756)	-30.043 (±10.460)		-3.305 (±8.837)	6.916 (±8.946)	23.118 (±6.894)	23.670 (±10.134)	27.351 (±7.850)	32.088** (±7.452)	28.685* (±7.331)
5	-94.451** (±11.662)	-55.413* (±14.037)	-26.738 (±11.853)	3.305 (±8.837)		10.221 (±8.938)	26.423 (±8.501)	26.975 (±9.947)	30.656 (±9.171)	35.394* (±9.028)	31.990 (±9.316)
6	-104.672** (±11.274)	-65.634** (±11.902)	-36.959 (±12.027)	-6.916 (±8.946)	-10.221 (±8.938)		16.202 (±6.918)	16.754 (±7.972)	20.435 (±6.338)	25.173** (±5.383)	21.769* (±5.623)
7	-120.875** (±12.091)	-81.836** (±11.853)	-53.161** (±10.511)	-23.118 (±6.894)	-26.423 (±8.501)	-16.202 (±6.918)		.551 (±6.602)	4.233 (±5.055)	8.970 (±4.395)	5.567 (±5.000)
8	-121.426** (±11.874)	-82.388** (±12.499)	-53.713* (±13.767)	-23.670 (±10.134)	-26.975 (±9.947)	-16.754 (±7.972)	-.551 (±6.602)		3.681 (±5.835)	8.419 (±5.538)	5.015 (±6.323)
9	-125.107** (±11.768)	-86.069** (±11.970)	-57.394** (±11.881)	-27.351 (±7.850)	-30.656 (±9.171)	-20.435 (±6.338)	-4.233 (±5.055)	-3.681 (±5.835)		4.738 (±3.819)	1.334 (±4.906)
10	-129.845** (±11.406)	-90.806** (±11.497)	-62.132** (±11.315)	-32.088** (±7.452)	-35.394* (±9.028)	-25.173** (±5.383)	-8.970 (±4.395)	-8.419 (±5.538)	-4.738 (±3.819)		-3.403 (±2.005)
11	-126.442** (±11.582)	-87.403** (±11.302)	-58.728** (±11.383)	-28.685* (±7.331)	-31.990 (±9.316)	-21.769* (±5.623)	-5.567 (±5.000)	-5.015 (±6.323)	-1.334 (±4.906)	3.403 (±2.005)	

**P<.01 * P<.05

Pairwise comparison table depicting results based on the results of a two-way ANOVA depicting the relationship between Tracking Accuracy and the nSweeps condition for the newborn and adult data at one iteration.

TA	1	2	3	4	5	6	7	8	9	10	11
1		.119** (±.026)	.161** (±.030)	.236** (±.028)	.295** (±.029)	.309** (±.034)	.369** (±.029)	.375** (±.032)	.382** (±.031)	.402** (±.031)	.409** (±.028)
2	-.119** (±.026)		.043 (±.027)	.118** (±.029)	.177** (±.029)	.190** (±.033)	.250** (±.029)	.256** (±.029)	.264** (±.029)	.284** (±.029)	.291** (±.026)
3	-.161** (±.030)	-.043 (±.027)		.075 (±.024)	.134** (±.032)	.147** (±.031)	.208** (±.027)	.213** (±.032)	.221** (±.026)	.241** (±.029)	.248** (±.027)
4	-.236** (±.028)	-.118** (±.029)	-.075 (±.024)		.059 (±.028)	.072 (±.029)	.132** (±.026)	.138** (±.031)	.146** (±.025)	.166** (±.027)	.173** (±.025)
5	-.295** (±.029)	-.177** (±.029)	-.134** (±.032)	-.059 (±.028)		.013 (±.029)	.074 (±.028)	.080* (±.022)	.087 (±.026)	.107* (±.028)	.114** (±.025)
6	-.309** (±.034)	-.190** (±.033)	-.147** (±.031)	-.072 (±.029)	-.013 (±.029)		.060 (±.022)	.066 (±.027)	.073 (±.021)	.094** (±.022)	.100** (±.021)
7	-.369** (±.029)	-.250** (±.029)	-.208** (±.027)	-.132** (±.026)	-.074 (±.028)	-.060 (±.022)		.006 (±.024)	.013 (±.018)	.033 (±.015)	.040 (±.017)
8	-.375** (±.032)	-.256** (±.029)	-.213** (±.032)	-.138** (±.031)	-.080* (±.022)	-.066 (±.027)	-.006 (±.024)		.007 (±.020)	.027 (±.019)	.034 (±.016)
9	-.382** (±.031)	-.264** (±.029)	-.221** (±.026)	-.146** (±.025)	-.087 (±.026)	-.073 (±.021)	-.013 (±.018)	-.007 (±.020)		.020 (±.013)	.027 (±.016)
10	-.402** (±.031)	-.284** (±.029)	-.241** (±.029)	-.166** (±.027)	-.107* (±.028)	-.094** (±.022)	-.033 (±.015)	-.027 (±.019)	-.020 (±.013)		.007 (±.011)
11	-.409** (±.028)	-.291** (±.026)	-.248** (±.027)	-.173** (±.025)	-.114 ** (±.025)	-.100** (±.021)	-.040 (±.017)	-.034 (±.016)	-.027 (±.016)	-.007 (±.011)	

**P<.01 * P<.05

Pairwise comparison table depicting results based on the results of a two-way ANOVA depicting the relationship between Frequency Error and the nSweeps condition for the adult data at one iteration.

FE	1	2	3	4	5	6	7	8	9	10	11
1		-1.581 ($\pm$.518)	-3.258** ($\pm$.614)	-4.318** ($\pm$.483)	-5.795** ($\pm$.507)	-6.104** ($\pm$.523)	-6.570** ($\pm$.472)	-6.771** ($\pm$.520)	-6.640** ($\pm$.531)	-6.841** ($\pm$.499)	-6.915** ($\pm$.491)
2	1.581 ($\pm$.518)		-1.677 ($\pm$.618)	-2.737** ($\pm$.609)	-4.214** ($\pm$.520)	-4.524** ($\pm$.609)	-4.989** ($\pm$.586)	-5.190** ($\pm$.603)	-5.059** ($\pm$.617)	-5.260** ($\pm$.615)	-5.334** ($\pm$.587)
3	3.258** ($\pm$.614)	1.677 ($\pm$.618)		-1.061 ($\pm$.516)	-2.537** ($\pm$.564)	-2.847** ($\pm$.630)	-3.312** ($\pm$.579)	-3.514** ($\pm$.633)	-3.383** ($\pm$.588)	-3.583** ($\pm$.615)	-3.657** ($\pm$.653)
4	4.318** ($\pm$.483)	2.737** ($\pm$.609)	1.061 ($\pm$.516)		-1.477* ($\pm$.378)	-1.786* ($\pm$.455)	-2.252** ($\pm$.459)	-2.453 ($\pm$.521)	-2.322** ($\pm$.435)	-2.522** ($\pm$.488)	-2.597** ($\pm$.471)
5	5.795** ($\pm$.507)	4.214** ($\pm$.520)	2.537** ($\pm$.564)	1.477* ($\pm$.378)		-.310 ($\pm$.402)	-.775** ($\pm$.407)	-.976 ($\pm$.427)	-.845 ($\pm$.416)	-1.046 ($\pm$.435)	-1.120 ($\pm$.360)
6	6.104** ($\pm$.523)	4.524** ($\pm$.609)	2.847** ($\pm$.630)	1.786* ($\pm$.455)	.310 ($\pm$.402)		-.466 ($\pm$.250)	-.667 ($\pm$.284)	-.536 ($\pm$.252)	-.736 ($\pm$.240)	-.810 ($\pm$.252)
7	6.570** ($\pm$.472)	4.989** ($\pm$.586)	3.312** ($\pm$.579)	2.252** ($\pm$.459)	.775 ($\pm$.407)	.466 ($\pm$.250)		-.201 ($\pm$.240)	-.070 ($\pm$.252)	-.271 ($\pm$.170)	-.345 ($\pm$.220)
8	6.771** ($\pm$.520)	5.190** ($\pm$.603)	3.514** ($\pm$.633)	2.453** ($\pm$.521)	.976 ($\pm$.427)	.667 ($\pm$.284)	.201 ($\pm$.240)		.131 ($\pm$.211)	-.069 ($\pm$.161)	-.144 ($\pm$.200)
9	6.640** ($\pm$.531)	5.059** ($\pm$.617)	3.383** ($\pm$.588)	2.322** ($\pm$.435)	.845 ($\pm$.416)	.536 ($\pm$.252)	.070 ($\pm$.252)	-.131 ($\pm$.211)		-.201 ($\pm$.179)	-.275 ($\pm$.282)
10	6.841** ($\pm$.499)	5.260** ($\pm$.615)	3.583** ($\pm$.615)	2.522** ($\pm$.488)	1.046 ($\pm$.435)	.736 ($\pm$.240)	.271 ($\pm$.170)	.069 ($\pm$.161)	.201 ($\pm$.179)		-.074 ($\pm$.170)
11	6.915** ($\pm$.491)	5.334** ($\pm$.587)	3.657** ($\pm$.653)	2.597** ($\pm$.471)	1.120 ($\pm$.360)	.810 ($\pm$.252)	.345 ($\pm$.220)	.144 ($\pm$.200)	.275 ($\pm$.282)	.074 ($\pm$.170)	

**P<.01 * P<.05

Pairwise comparison table depicting results based on the results of a two-way ANOVA depicting the relationship between Slope Error and the nSweeps condition for the adult data at one iteration.

SE	1	2	3	4	5	6	7	8	9	10	11
1		32.667 (±16.657)	65.475 (±19.495)	81.908** (±17.548)	82.998** (±16.871)	97.499** (±15.005)	108.077** (±17.299)	115.048** (±15.336)	113.071** (±14.524)	111.652** (±16.402)	107.145** (±16.748)
2	-32.667 (±16.657)		32.808 (±13.067)	49.241 (±14.300)	50.331 (±14.702)	64.832* (±15.228)	75.410** (±16.029)	82.382** (±17.234)	80.404** (±15.376)	78.986** (±16.082)	74.478** (±15.586)
3	-65.475 (±19.495)	-32.808 (±13.067)		16.433 (±11.299)	17.523 (±10.526)	32.024 (±12.906)	42.602* (±11.202)	49.574 (±14.533)	47.596 (±13.515)	46.177* (±11.970)	41.670 (±11.573)
4	-81.908** (±17.548)	-49.241 (±14.300)	-16.433 (±11.299)		1.090 (±8.730)	15.591 (±10.875)	26.169 (±9.661)	33.141 (±14.074)	31.163 (±11.733)	29.744 (±10.246)	25.237 (±8.944)
5	-82.998** (±16.871)	-50.331 (±14.702)	-17.523 (±10.526)	-1.090 (±8.730)		14.501 (±10.061)	25.079 (±10.297)	32.050 (±12.319)	30.073 (±10.292)	28.654 (±9.810)	24.147 (±9.173)
6	-97.499** (±15.005)	-64.832* (±15.228)	-32.024 (±12.906)	-15.591 (±10.875)	-14.501 (±10.061)		10.578 (±6.677)	17.550 (±8.149)	15.572 (±5.985)	14.153 (±4.806)	9.646 (±5.146)
7	- 108.077** (±17.299)	-75.410** (±16.029)	-42.602* (±11.202)	-26.169 (±9.661)	-25.079 (±10.297)	-10.578 (±6.677)		6.972 (±7.397)	4.994 (±6.798)	3.575 (±4.403)	-.932 (±5.119)
8	- 115.048** (±15.336)	-82.382** (±17.234)	-49.574 (±14.533)	-33.141 (±14.074)	-32.050 (±12.319)	-17.550 (±8.149)	-6.972 (±7.397)		-1.978 (±4.938)	-3.396 (±7.131)	-7.903 (±9.134)
9	- 113.071** (±14.524)	-80.404** (±15.376)	-47.596 (±13.515)	-31.163 (±11.733)	-30.073 (±10.292)	-15.572 (±5.985)	-4.994 (±6.798)	1.978 (±4.938)		-1.418 (±4.572)	-5.925 (±5.887)
10	- 111.652** (±16.402)	-78.986** (±16.082)	-46.177* (±11.970)	-29.744 (±10.246)	-28.654 (±9.810)	-14.153 (±4.806)	-3.575 (±4.403)	3.396 (±7.131)	1.418 (±4.572)		-4.507 (±2.629)
11	- 107.145** (±16.748)	-74.478** (±15.586)	-41.670 (±11.573)	-25.237 (±8.944)	-24.147 (±9.173)	-9.646 (±5.146)	.932 (±5.119)	7.903 (±9.134)	5.925 (±5.887)	4.507 (±2.629)	

**P<.01 * P<.05

Pairwise comparison table depicting results based on the results of a two-way ANOVA depicting the relationship between Tracking Accuracy and the nSweeps condition for the adult data at one iteration.

TA	1	2	3	4	5	6	7	8	9	10	11
1		.103 (±.030)	.187 (±.035)	.246** (±.031)	.296** (±.028)	.306** (±.038)	.335** (±.040)	.358** (±.037)	.347** (±.039)	.344** (±.043)	.359** (±.034)
2	-.103 (±.030)		.083 (±.036)	.143* (±.034)	.192** (±.031)	.203** (±.043)	.232** (±.042)	.255** (±.038)	.244** (±.038)	.241** (±.044)	.255** (±.035)
3	-.187** (±.035)	-.083 (±.036)		.059 (±.029)	.109* (±.027)	.119 (±.041)	.148* (±.036)	.171** (±.037)	.161** (±.037)	.158* (±.040)	.172** (±.036)
4	-.246** (±.031)	-.143* (±.034)	-.059 (±.029)		.049 (±.023)	.060 (±.035)	.089 (±.035)	.112 (±.037)	.101 (±.031)	.098 (±.037)	.112 (±.031)
5	-.296** (±.028)	-.192** (±.031)	-.109* (±.027)	-.049 (±.023)		.011 (±.035)	.040 (±.037)	.063 (±.029)	.052 (±.030)	.049 (±.038)	.063 (±.027)
6	-.306** (±.038)	-.203** (±.043)	-.119 (±.041)	-.060 (±.035)	-.011 (±.035)		.029 (±.018)	.052 (±.026)	.041 (±.021)	.038 (±.016)	.052 (±.018)
7	-.335** (±.040)	-.232** (±.042)	-.148* (±.036)	-.089 (±.035)	-.040 (±.037)	-.029 (±.018)		.023 (±.026)	.012 (±.020)	.009 (±.010)	.023 (±.020)
8	-.358** (±.037)	-.255** (±.038)	-.171** (±.037)	-.112 (±.037)	-.063 (±.029)	-.052 (±.026)	-.023 (±.026)		-.011 (±.022)	-.014 (±.026)	.000 (±.014)
9	-.347** (±.039)	-.244** (±.038)	-.161** (±.037)	-.101 (±.031)	-.052 (±.030)	-.041 (±.021)	-.012 (±.020)	.011 (±.022)		-.003 (±.015)	.011 (±.019)
10	-.344** (±.043)	-.241** (±.044)	-.158* (±.040)	-.098 (±.037)	-.049 (±.038)	-.038 (±.016)	-.009 (±.010)	.014 (±.026)	.003 (±.015)		.014 (±.021)
11	-.359** (±.034)	-.255** (±.035)	-.172** (±.036)	-.112 (±.031)	-.063 (±.027)	-.052 (±.018)	-.023 (±.020)	.000 (±.014)	-.011 (±.019)	-.014 (±.021)	

**P<.01 * P<.05

Pairwise comparison table depicting results based on the results of a two-way ANOVA depicting the relationship between Frequency Error and the nSweeps condition for the newborn data at one iteration.

FE	1	2	3	4	5	6	7	8	9	10	11
1		-2.267 (±.816)	-1.978 (±.855)	-3.404* (±.845)	-5.084** (±.720)	-5.844** (±.802)	-6.730** (±.697)	-7.034** (±.756)	-7.083** (±.762)	-7.598** (±.764)	-7.622** (±.740)
2	2.267 (±.816)		.289 (±.657)	-1.137 (±.797)	-2.817* (±.705)	-3.577** (±.790)	-4.462** (±.686)	-4.767** (±.673)	-4.816** (±.682)	-5.331** (±.708)	-5.355** (±.763)
3	1.978 (±.855)	-.289 (±.657)		-1.425 (±.597)	-3.106* (±.788)	-3.865** (±.779)	-4.751** (±.773)	-5.056** (±.732)	-5.105** (±.702)	-5.620** (±.745)	-5.643** (±.790)
4	3.404* (±.845)	1.137 (±.797)	1.425 (±.597)		-1.681* (±.615)	-2.440* (±.645)	-3.326** (±.716)	-3.631** (±.758)	-3.679** (±.732)	-4.194** (±.750)	-4.218** (±.772)
5	5.084** (±.720)	2.817* (±.705)	3.106* (±.788)	1.681 (±.615)		-.759 (±.482)	-1.645 (±.564)	-1.950** (±.447)	-1.999* (±.501)	-2.514** (±.537)	-2.537** (±.553)
6	5.844** (±.802)	3.577** (±.790)	3.865** (±.779)	2.440* (±.645)	.759 (±.482)		-.886 (±.498)	-1.191 (±.544)	-1.239 (±.534)	-1.754 (±.554)	-1.778 (±.526)
7	6.730** (±.697)	4.462** (±.686)	4.751** (±.773)	3.326** (±.716)	1.645 (±.564)	.886 (±.498)		-.305 (±.472)	-.354 (±.458)	-.869 (±.477)	-.892 (±.457)
8	7.034** (±.756)	4.767** (±.673)	5.056** (±.732)	3.631** (±.758)	1.950** (±.447)	1.191 (±.544)	.305 (±.472)		-.049 (±.239)	-.564 (±.270)	-.587 (±.260)
9	7.083** (±.762)	4.816** (±.682)	5.105** (±.702)	3.679** (±.732)	1.999* (±.501)	1.239 (±.534)	.354 (±.458)	.049 (±.239)		-.515 (±.172)	-.539 (±.235)
10	7.598** (±.764)	5.331** (±.708)	5.620** (±.745)	4.194** (±.750)	2.514* (±.537)	1.754 (±.554)	.869 (±.477)	.564 (±.270)	.515 (±.172)		-.024 (±.164)
11	7.622** (±.740)	5.355** (±.763)	5.643** (±.790)	4.218** (±.772)	2.537** (±.553)	1.778 (±.526)	.892 (±.457)	.587 (±.260)	.539 (±.235)	.024 (±.164)	

**P<.01 * P<.05

Pairwise comparison table depicting results based on the results of a two-way ANOVA depicting the relationship between Slope Error and the nSweeps condition for the newborn data at one iteration.

SE	1	2	3	4	5	6	7	8	9	10	11
1		45.411 (±22.242)	69.952 (±21.758)	113.605** (±19.149)	105.905** (±16.107)	111.846** (±16.830)	133.673** (±16.898)	127.804** (±18.133)	137.144** (±18.521)	148.038** (±15.853)	145.738** (±16.003)
2	-45.411 (±22.242)		24.541 (±18.962)	68.195 (±18.662)	60.494 (±23.918)	66.435 (±18.295)	88.262** (±17.467)	82.393** (±18.107)	91.734** (±18.349)	102.627** (±16.435)	100.327** (±16.370)
3	-69.952 (±21.758)	-24.541 (±18.962)		43.654 (±17.606)	35.953 (±21.242)	41.894 (±20.298)	63.721 (±17.790)	57.852 (±23.386)	67.192 (±19.544)	78.086* (±19.206)	75.786* (±19.605)
4	-113.605 (±19.149)	-68.195 (±18.662)	-43.654 (±17.606)		-7.701 (±15.367)	-1.760 (±14.208)	20.067 (±9.836)	14.199 (±14.586)	23.539 (±10.433)	34.432 (±10.824)	32.132 (±11.618)
5	- 105.905** (±16.107)	-60.494 (±23.918)	-35.953 (±21.242)	7.701 (±15.367)		5.941 (±14.777)	27.768 (±13.529)	21.899 (±15.622)	31.239 (±15.183)	42.133 (±15.159)	39.833 (±16.217)
6	- 111.846** (±16.830)	-66.435 (±18.295)	-41.894 (±20.298)	1.760 (±14.208)	-5.941 (±14.777)		21.827 (±12.119)	15.958 (±13.704)	25.298 (±11.175)	36.192* (±9.634)	33.892 (±10.000)
7	- 133.673** (±16.898)	-88.262** (±17.467)	-63.721 (±17.790)	-20.067 (±9.836)	-27.768 (±15.622)	-21.827 (±12.119)		-5.869 (±10.937)	3.472 (±7.483)	14.365 (±7.608)	12.065 (±8.591)
8	- 127.804** (±18.133)	-82.393** (±18.107)	-57.852 (±23.386)	-14.199 (±14.586)	-21.899 (±13.529)	-15.958 (±13.704)	5.869 (±10.937)		9.340 (±10.573)	20.234 (±8.475)	17.934 (±8.746)
9	- 137.144** (±18.521)	-91.734** (±18.349)	-67.192 (±19.544)	-23.539 (±10.433)	-31.239 (±15.183)	-25.298 (±11.175)	-3.472 (±7.483)	-9.340 (±10.573)		10.893 (±6.119)	8.594 (±7.851)
10	- 148.038** (±15.853)	102.627** (±16.435)	-78.086* (±19.206)	-34.432 (±10.824)	-42.133 (±15.159)	-36.192* (±9.634)	-14.365 (±7.608)	-20.234 (±8.475)	-10.893 (±6.119)		-2.300 (±3.029)
11	- 145.738** (±16.003)	- 100.327** (±16.370)	-75.786* (±19.605)	-32.132 (±11.618)	-39.833 (±16.217)	-33.892 (±10.000)	-12.065 (±8.591)	-17.934 (±8.746)	-8.594 (±7.851)	2.300 (±3.029)	

**P<.01 * P<.05

Pairwise comparison table depicting results based on the results of a two-way ANOVA depicting the relationship between Tracking Accuracy and the nSweeps condition for the newborn data at one iteration.

TA	1	2	3	4	5	6	7	8	9	10	11
1		.134 (±.041)	.136 (±.049)	.227** (±.048)	.295** (±.050)	.311** (±.056)	.403** (±.042)	.391** (±.052)	.417** (±.049)	.460** (±.045)	.460** (±.045)
2	-.134 (±.041)		.002 (±.041)	.093 (±.046)	.161 (±.050)	.178 (±.051)	.269** (±.040)	.258** (±.044)	.283** (±.043)	.327** (±.038)	.326** (±.038)
3	-.136 (±.049)	-.002 (±.041)		.091 (±.039)	.159 (±.058)	.175* (±.047)	.267** (±.040)	.256** (±.053)	.281** (±.037)	.324** (±.041)	.324** (±.040)
4	-.227** (±.048)	-.093 (±.046)	-.091 (±.039)		.068 (±.051)	.084 (±.045)	.176** (±.038)	.165 (±.050)	.190** (±.039)	.233** (±.039)	.233** (±.039)
5	-.295** (±.050)	-.161 (±.050)	-.159 (±.058)	-.068 (±.051)		.016 (±.047)	.108 (±.042)	.096 (±.034)	.122 (±.042)	.165* (±.040)	.165* (±.041)
6	-.311** (±.056)	-.178 (±.051)	-.175* (±.047)	-.084 (±.045)	-.016 (±.047)		.092 (±.040)	.080 (±.047)	.105 (±.036)	.149* (±.040)	.148* (±.039)
7	-.403** (±.042)	-.269** (±.040)	-.267** (±.040)	-.176** (±.038)	-.108 (±.042)	-.092 (±.040)		-.011 (±.040)	.014 (±.030)	.058 (±.029)	.057 (±.029)
8	-.391** (±.052)	-.258** (±.044)	-.256** (±.053)	-.165 (±.050)	-.096 (±.034)	-.080 (±.047)	.011 (±.040)		.025 (±.032)	.069 (±.028)	.068 (±.028)
9	-.417** (±.049)	-.283** (±.043)	-.281** (±.037)	-.190** (±.039)	-.122 (±.042)	-.105 (±.036)	-.014 (±.030)	-.025 (±.032)		.044 (±.021)	.043 (±.025)
10	-.460** (±.045)	-.327** (±.038)	-.324** (±.041)	-.233** (±.039)	-.165* (±.040)	-.149* (±.040)	-.058 (±.029)	-.069 (±.028)	-.044 (±.021)		-.001 (±.008)
11	-.460** (±.045)	-.326** (±.038)	-.324** (±.040)	-.233** (±.039)	-.165* (±.041)	-.148* (±.039)	-.057 (±.029)	-.068 (±.028)	-.043 (±.025)	.001 (±.008)	

**P<.01 * P<.05

Appendix D

5 Iteration nSweeps Confusion Matrices

Pairwise comparison table depicting results based on the results of a one-way ANOVA depicting the relationship between Frequency Error and the nSweeps condition for the hidden component of newborn and adult data at five iterations.

FE	1	2	3	4	5	6	7	8	9	10	11
1		.358 (±.329)	.970 (±.424)	2.181** (±.433)	3.680** (±.429)	4.015** (±.445)	4.456** (±.408)	4.785** (±.417)	5.073** (±.428)	5.367** (±.389)	5.436** (±.398)
2	-.358 (±.329)		.612 (±.367)	1.824** (±.362)	3.322** (±.390)	3.657** (±.395)	4.098** (±.377)	4.427** (±.359)	4.715** (±.368)	5.009** (±.338)	5.078** (±.349)
3	-.970 (±.424)	-.612 (±.367)		1.212 (±.349)	2.710** (±.363)	3.045** (±.372)	3.486** (±.358)	3.815** (±.366)	4.103** (±.366)	4.397** (±.363)	4.466** (±.372)
4	-2.181** (±.433)	-1.824** (±.362)	-1.212 (±.349)		1.499** (±.302)	1.834** (±.351)	2.275** (±.337)	2.603** (±.337)	2.892** (±.348)	3.186** (±.325)	3.254** (±.337)
5	-3.680** (±.429)	-3.322** (±.390)	-2.710** (±.363)	-1.499** (±.302)		.335 (±.248)	.776 (±.230)	1.105** (±.258)	1.393** (±.285)	1.687** (±.241)	1.756** (±.254)
6	-4.015** (±.445)	-3.657** (±.395)	-3.045** (±.372)	-1.834** (±.351)	-.335 (±.248)		.441 (±.158)	.770** (±.175)	1.058** (±.206)	1.352** (±.171)	1.421** (±.173)
7	-4.456** (±.408)	-4.098** (±.377)	-3.486** (±.358)	-2.275** (±.337)	-.776 (±.230)	-.441 (±.158)		.329 (±.135)	.617 (±.177)	.911** (±.151)	.980** (±.155)
8	-4.785** (±.417)	-4.427** (±.359)	-3.815** (±.366)	-2.603** (±.337)	-1.105** (±.258)	-.770** (±.175)	-.329 (±.135)		.288 (±.121)	.582** (±.117)	.651** (±.115)
9	-5.073** (±.428)	-4.715** (±.363)	-4.103** (±.366)	-2.892** (±.348)	-1.393** (±.285)	-1.058** (±.206)	-.617 (±.177)	-.288 (±.121)		.294 (±.115)	.363 (±.111)
10	-5.367** (±.389)	-5.009** (±.338)	-4.397** (±.363)	-3.186** (±.325)	-1.687** (±.241)	-1.352** (±.171)	-.911** (±.151)	-.582** (±.117)	-.294 (±.115)		.069 (±.059)
11	-5.436** (±.398)	-5.073** (±.349)	-4.466** (±.372)	-3.254 ** (±.337)	-1.756** (±.254)	-1.421** (±.173)	-.980 ** (±.155)	-.651 ** (±.115)	-.363 (±.111)	-.069 (±.059)	

**P<.01 * P<.05

Pairwise comparison table depicting results based on the results of a one-way ANOVA depicting the relationship between Slope Error and the nSweeps condition for the hidden component of newborn and adult data at five iterations.

SE	1	2	3	4	5	6	7	8	9	10	11
1		-28.130* (±7.928)	-44.272** (±8.188)	-61.612** (±8.539)	-88.674** (±8.176)	-97.757** (±8.590)	-105.725** (±7.623)	-107.732** (±8.586)	-114.181** (±7.997)	-115.293** (±8.503)	-119.154** (±8.115)
2	28.130* (±7.928)		-16.142 (±6.507)	-33.482** (±6.335)	-60.544** (±6.752)	-69.626** (±7.612)	-77.595** (±6.762)	-79.602** (±7.291)	-86.051** (±7.135)	-87.163** (±7.695)	-91.024** (±7.342)
3	44.272** (±8.188)	16.142 (±6.507)		-17.341 (±6.875)	-44.402** (±6.296)	-53.485** (±7.077)	-61.453** (±6.500)	-63.461** (±7.378)	-69.909** (±7.120)	-71.021** (±7.535)	-74.883** (±7.172)
4	61.612** (±8.539)	33.482** (±6.335)	17.341 (±6.875)		-27.061** (±5.149)	-36.144** (±6.208)	-44.112** (±5.053)	-46.120** (±5.619)	-52.569** (±5.231)	-53.680** (±5.522)	-57.542** (±5.503)
5	88.674** (±8.176)	60.544** (±6.752)	44.402** (±6.296)	27.061** (±5.149)		-9.083 (±5.449)	-17.051* (±4.365)	-19.059* (±4.884)	-25.507** (±4.931)	-26.619** (±4.955)	-30.481** (±5.019)
6	97.757** (±8.590)	69.626** (±7.612)	53.485** (±7.077)	36.144** (±6.208)	9.083 (±5.449)		-7.968 (±4.164)	-9.976 (±4.864)	-16.425* (±4.409)	-17.536* (±4.836)	-21.398** (±4.727)
7	105.725** (±7.623)	77.595** (±6.762)	61.453** (±6.500)	44.112** (±5.053)	17.051* (±4.365)	7.968 (±4.164)		-2.008 (±2.961)	-8.456 (±3.134)	-9.568 (±3.494)	-13.430** (±3.026)
8	107.732** (±8.586)	79.602** (±7.291)	63.461** (±7.378)	46.120** (±5.619)	19.059* (±4.884)	9.976 (±4.864)	2.008 (±2.961)		-6.449 (±2.740)	-7.560* (±2.110)	-11.422** (±2.303)
9	114.181** (±7.997)	86.051** (±7.135)	69.909** (±7.120)	52.569** (±5.231)	25.507** (±4.931)	16.425* (±4.409)	8.456 (±3.134)	6.449 (±2.740)		-1.112 (±2.110)	-4.973 (±1.694)
10	115.293** (±8.503)	87.163** (±7.695)	71.021** (±7.535)	53.680** (±5.522)	26.619** (±4.955)	17.536* (±4.836)	9.568 (±3.494)	7.560* (±2.110)	1.112 (±2.110)		-3.862 (±1.442)
11	119.154** (±8.115)	91.024** (±7.342)	74.883** (±7.172)	57.542** (±5.503)	30.481** (±5.019)	21.398** (±4.727)	13.430** (±3.026)	11.422** (±2.303)	4.973 (±1.694)	3.862 (±1.442)	

**P<.01 * P<.05

Pairwise comparison table depicting results based on the results of a one-way ANOVA depicting the relationship between Tracking Accuracy and the nSweeps condition for the hidden component of newborn and adult data at five iterations.

TA	1	2	3	4	5	6	7	8	9	10	11
1		-.053 (±.027)	-.071 (±.036)	-.118 (±.035)	-.177** (±.030)	-.191** (±.032)	-.221** (±.030)	-.239** (±.031)	-.251** (±.032)	-.269** (±.031)	-.281** (±.030)
2	.053 (±.027)		-.018 (±.029)	-.065 (±.029)	-.124** (±.028)	-.137** (±.031)	-.168** (±.026)	-.186** (±.027)	-.198** (±.029)	-.215** (±.026)	-.228** (±.025)
3	.071 (±.036)	.018 (±.029)		-.047 (±.024)	-.105* (±.029)	-.119** (±.029)	-.150** (±.026)	-.167** (±.026)	-.180** (±.026)	-.197** (±.027)	-.209** (±.026)
4	.118 (±.035)	.065 (±.029)	.047 (±.024)		-.059 (±.025)	-.072 (±.029)	-.103** (±.025)	-.121** (±.024)	-.133** (±.025)	-.150** (±.026)	-.163** (±.025)
5	.177** (±.030)	.124** (±.028)	.105* (±.029)	.059 (±.025)		-.014 (±.020)	-.044 (±.020)	-.062 (±.021)	-.074 (±.024)	-.092** (±.022)	-.104** (±.022)
6	.191** (±.032)	.137** (±.031)	.119** (±.029)	.072 (±.029)	.014 (±.020)		-.031 (±.015)	-.048 (±.017)	-.061 (±.020)	-.078** (±.019)	-.090** (±.018)
7	.221** (±.030)	.168** (±.026)	.150** (±.026)	.103** (±.025)	.044 (±.020)	.031 (±.015)		-.017 (±.010)	-.030 (±.014)	-.047* (±.013)	-.060** (±.012)
8	.239** (±.031)	.186** (±.027)	.167** (±.026)	.121** (±.024)	.062 (±.021)	.048 (±.017)	.017 (±.010)		-.012 (±.011)	-.030 (±.010)	-.042** (±.009)
9	.251** (±.032)	.198** (±.029)	.180** (±.026)	.133** (±.025)	.074 (±.024)	.061 (±.020)	.030 (±.014)	.012 (±.011)		-.017 (±.007)	-.030* (±.008)
10	.269** (±.031)	.215** (±.026)	.197** (±.027)	.150** (±.026)	.092** (±.022)	.078** (±.019)	.047* (±.013)	.030 (±.010)	.017 (±.007)		-.012 (±.005)
11	.281** (±.030)	.228** (±.025)	.209** (±.026)	.163** (±.025)	.104** (±.022)	.090** (±.018)	.060 ** (±.012)	.042** (±.009)	.030 (±.008)	.012 (±.005)	

**P<.01 * P<.05

Pairwise comparison table depicting results based on the results of a one-way ANOVA depicting the relationship between Frequency Error and the nSweeps condition for the hidden component of the adult data at five iterations.

FE	1	2	3	4	5	6	7	8	9	10	11
1		.478 (±.471)	1.581 (±.692)	3.099** (±.693)	4.420** (±.666)	4.938** (±.698)	5.272** (±.649)	5.252** (±.699)	5.464** (±.699)	5.619** (±.652)	5.623** (±.664)
2	-.478 (±.471)		1.103 (±.487)	2.621** (±.509)	3.943** (±.456)	4.460** (±.476)	4.794** (±.451)	4.774** (±.514)	4.986** (±.533)	5.142** (±.467)	5.145** (±.487)
3	-1.581 (±.692)	-1.103 (±.487)		1.518 (±.471)	2.839** (±.411)	3.357** (±.502)	3.691** (±.503)	3.671** (±.567)	3.883** (±.570)	4.038** (±.566)	4.042** (±.575)
4	-3.099** (±.693)	-2.621** (±.509)	-1.518 (±.471)		1.322* (±.332)	1.839** (±.350)	2.173** (±.386)	2.153** (±.437)	2.365** (±.457)	2.521** (±.426)	2.524** (±.447)
5	-4.420** (±.666)	-3.943** (±.456)	-2.839** (±.411)	-1.322* (±.332)		.518 (±.239)	.852* (±.225)	.831 (±.300)	1.043 (±.292)	1.199** (±.273)	1.202* (±.285)
6	-4.938** (±.698)	-4.460** (±.476)	-3.357** (±.502)	-1.839** (±.350)	-.518 (±.239)		.334 (±.179)	.314 (±.196)	.526 (±.205)	.681* (±.173)	.685* (±.185)
7	-5.272** (±.649)	-4.794** (±.451)	-3.691** (±.503)	-2.173** (±.386)	-.852* (±.225)	-.334 (±.179)		-.020 (±.165)	.192 (±.180)	.347 (±.137)	.351 (±.147)
8	-5.252** (±.699)	-4.774** (±.514)	-3.671** (±.567)	-2.153** (±.437)	-.831 (±.300)	-.314 (±.196)	.020 (±.165)		.212 (±.113)	.367 (±.125)	.371 (±.122)
9	-5.464** (±.699)	-4.986** (±.533)	-3.883** (±.570)	-2.365** (±.457)	-1.043 (±.292)	-.526 (±.205)	-.192 (±.180)	-.212 (±.113)		.155 (±.133)	.159 (±.107)
10	-5.619** (±.652)	-5.142** (±.467)	-4.038** (±.566)	-2.521** (±.426)	-1.199** (±.273)	-.681* (±.173)	-.347 (±.137)	-.367 (±.125)	-.155 (±.133)		.003 (±.070)
11	-5.623** (±.664)	-5.145** (±.487)	-4.042** (±.575)	-2.524** (±.447)	-1.202* (±.285)	-.685* (±.185)	-.351 (±.147)	-.371 (±.122)	-.159 (±.107)	-.003 (±.070)	

**P<.01 * P<.05

Pairwise comparison table depicting results based on the results of a one-way ANOVA depicting the relationship between Slope Error and the nSweeps condition for the hidden component of the adult data at five iterations.

SE	1	2	3	4	5	6	7	8	9	10	11
1		-31.833 (±11.397)	-51.288* (±12.182)	- 79.278** (±12.266)	- 95.703** (±12.053)	- 108.214** (±12.882)	- 106.877** (±11.872)	- 110.926** (±12.890)	- 113.498** (±12.174)	- 115.469** (±12.404)	- 116.579** (±12.185)
2	31.833 (±11.397)		-19.454 (±7.516)	- 47.445** (±7.648)	- 63.870** (±9.142)	-76.381** (±9.111)	-75.043** (±9.286)	-79.092** (±9.787)	-81.665** (±9.639)	-83.635** (±9.787)	-84.745** (±9.840)
3	51.288* (±12.182)	19.454 (±7.516)		- 27.990** (±6.332)	- 44.415** (±7.975)	-56.926** (±8.450)	-55.589** (±8.717)	-59.638** (±9.257)	-62.210** (±9.293)	-64.181** (±9.395)	-65.291** (±9.475)
4	79.278** (±12.266)	47.445** (±7.648)	27.990** (±6.332)		-16.425 (±6.389)	-28.936** (±6.494)	-27.599** (±5.883)	-31.648** (±6.073)	-34.220** (±6.708)	-36.191** (±6.523)	-37.301** (±7.119)
5	95.703** (±12.053)	63.870** (±9.142)	44.415** (±7.975)	16.425 (±6.389)		-12.511 (±5.255)	-11.174 (±5.416)	-15.223 (±5.307)	-17.795 (±6.492)	-19.766 (±5.822)	-20.876 (±6.182)
6	108.214** (±12.882)	76.381** (±9.111)	56.926** (±8.450)	28.936** (±6.494)	12.511 (±5.255)		1.338 (±4.660)	-2.711 (±3.954)	-5.284 (±4.804)	-7.254 (±4.880)	-8.364 (±5.159)
7	106.877** (±11.872)	75.043** (±9.286)	55.589** (±8.717)	27.599** (±5.883)	11.174 (±5.416)	-1.338 (±4.660)		-4.049 (±2.714)	-6.622 (±3.294)	-8.592 (±3.273)	-9.702 (±3.007)
8	110.926** (±12.890)	79.092** (±9.787)	59.638** (±9.257)	31.648** (±6.073)	15.223 (±5.307)	2.711 (±3.954)	4.049 (±2.714)		-2.573 (±2.894)	-4.543 (±2.212)	-5.653 (±2.783)
9	113.498** (±12.174)	81.665** (±9.639)	62.210** (±9.293)	34.220** (±6.708)	17.795 (±6.492)	5.284 (±4.804)	6.622 (±3.294)	2.573 (±2.894)		-1.970 (±2.686)	-3.080 (±2.037)
10	115.469** (±12.404)	83.635** (±9.787)	64.181** (±9.395)	36.191** (±6.523)	19.766 (±5.822)	7.254 (±4.880)	8.592 (±3.273)	4.543 (±2.212)	1.970 (±2.686)		-1.110 (±1.533)
11	116.579** (±12.185)	84.745** (±9.840)	65.291** (±9.475)	37.301** (±7.119)	20.876 (±6.182)	8.364 (±5.159)	9.702 (±3.007)	5.653 (±2.783)	3.080 (±2.037)	1.110 (±1.533)	

**P<.01 * P<.05

Pairwise comparison table depicting results based on the results of a one-way ANOVA depicting the relationship between Tracking Accuracy and the nSweeps condition for the hidden component of the adult data at five iterations.

TA	1	2	3	4	5	6	7	8	9	10	11
1		.018 (±.039)	-.035 (±.059)	-.115 (±.053)	-.196* (±.050)	-.228** (±.050)	-.234** (±.047)	-.236** (±.050)	-.245** (±.050)	-.260** (±.047)	-.263** (±.046)
2	-.018 (±.039)		-.053 (±.038)	-.133* (±.034)	-.214** (±.031)	-.246** (±.034)	-.251** (±.031)	-.253** (±.036)	-.263** (±.038)	-.278** (±.033)	-.281** (±.033)
3	.035 (±.059)	.053 (±.038)		-.080 (±.028)	-.161** (±.030)	-.193** (±.032)	-.199** (±.036)	-.201** (±.038)	-.210** (±.038)	-.225** (±.038)	-.228** (±.037)
4	.115 (±.053)	.133 (±.034)	.080 (±.028)		-.081* (±.021)	-.113** (±.021)	-.119** (±.023)	-.121** (±.027)	-.130** (±.030)	-.145** (±.028)	-.148** (±.028)
5	.196* (±.050)	.214** (±.031)	.161** (±.030)	.081* (±.021)		-.032 (±.016)	-.038 (±.018)	-.040 (±.023)	-.049 (±.026)	-.064 (±.021)	-.067 (±.022)
6	.228** (±.050)	.246** (±.034)	.193** (±.032)	.113** (±.021)	.032 (±.016)		-.005 (±.013)	-.007 (±.016)	-.017 (±.020)	-.032 (±.016)	-.035 (±.015)
7	.234** (±.047)	.251** (±.031)	.199** (±.036)	.119** (±.023)	.038 (±.018)	.005 (±.013)		-.002 (±.013)	-.012 (±.018)	-.026 (±.013)	-.030 (±.012)
8	.236** (±.050)	.253** (±.036)	.201** (±.038)	.121** (±.027)	.040 (±.023)	.007 (±.016)	.002 (±.013)		-.010 (±.011)	-.025 (±.008)	-.028 (±.008)
9	.245** (±.050)	.263** (±.038)	.210** (±.038)	.130** (±.030)	.049 (±.026)	.017 (±.020)	.012 (±.018)	.010 (±.011)		-.015 (±.010)	-.018 (±.010)
10	.260** (±.047)	.278** (±.033)	.225** (±.038)	.145** (±.028)	.064 (±.021)	.032 (±.016)	.026 (±.013)	.025 (±.008)	.015 (±.010)		-.003 (±.005)
11	.263** (±.046)	.281** (±.033)	.228** (±.037)	.148** (±.028)	.067 (±.022)	.035 (±.015)	.030 (±.012)	.028 (±.008)	.018 (±.010)	.003 (±.005)	

**P<.01 * P<.05

Pairwise comparison table depicting results based on the results of a one-way ANOVA depicting the relationship between Frequency Error and the nSweeps condition for the hidden component of the newborn data at five iterations.

FE	1	2	3	4	5	6	7	8	9	10	11
1		.238 ($\pm$.458)	.358 ($\pm$.481)	1.264 ($\pm$.512)	2.940** ($\pm$.534)	3.092** ($\pm$.546)	3.640** ($\pm$.488)	4.318** ($\pm$.444)	4.682** ($\pm$.485)	5.115** ($\pm$.415)	5.249** ($\pm$.429)
2	-.238 ($\pm$.458)		.120 ($\pm$.551)	1.026 ($\pm$.513)	2.702* ($\pm$.639)	2.854** ($\pm$.636)	3.402** ($\pm$.608)	4.080** ($\pm$.501)	4.444** ($\pm$.504)	4.877** ($\pm$.488)	5.011** ($\pm$.500)
3	-.358 ($\pm$.481)	-.120 ($\pm$.551)		.906 ($\pm$.516)	2.581* ($\pm$.605)	2.734** ($\pm$.551)	3.282** ($\pm$.510)	3.959** ($\pm$.459)	4.324** ($\pm$.453)	4.756** ($\pm$.449)	4.890** ($\pm$.466)
4	-1.264 ($\pm$.512)	-1.026 ($\pm$.513)	-.906 ($\pm$.516)		1.676 ($\pm$.510)	1.828 ($\pm$.616)	2.376* ($\pm$.558)	3.054** ($\pm$.514)	3.418** ($\pm$.526)	3.851** ($\pm$.494)	3.985** ($\pm$.507)
5	-2.940** ($\pm$.534)	-2.702* ($\pm$.639)	-2.581* ($\pm$.605)	-1.676 ($\pm$.510)		.152 ($\pm$.441)	.700 ($\pm$.406)	1.378 ($\pm$.422)	1.742 ($\pm$.495)	2.175** ($\pm$.402)	2.309** ($\pm$.425)
6	-3.092** ($\pm$.546)	-2.854** ($\pm$.636)	-2.734** ($\pm$.551)	-1.828 ($\pm$.616)	-.152 ($\pm$.441)		.548 ($\pm$.262)	1.226* ($\pm$.293)	1.590** ($\pm$.362)	2.023** ($\pm$.299)	2.157** ($\pm$.296)
7	-3.640** ($\pm$.488)	-3.402** ($\pm$.608)	-3.282** ($\pm$.510)	-2.376* ($\pm$.558)	-.700 ($\pm$.406)	-.548 ($\pm$.262)		.677 ($\pm$.215)	1.042 ($\pm$.309)	1.474** ($\pm$.272)	1.608** ($\pm$.276)
8	-4.318** ($\pm$.444)	-4.080** ($\pm$.501)	-3.959** ($\pm$.459)	-3.054** ($\pm$.514)	-1.378 ($\pm$.422)	-1.226* ($\pm$.293)	-.677 ($\pm$.215)		.365 ($\pm$.217)	.797* ($\pm$.199)	.931** ($\pm$.196)
9	-4.682** ($\pm$.485)	-4.444** ($\pm$.504)	-4.324** ($\pm$.453)	-3.418** ($\pm$.526)	-1.742 ($\pm$.495)	-1.590** ($\pm$.362)	-1.042 ($\pm$.309)	-.365 ($\pm$.217)		.432 ($\pm$.190)	.566 ($\pm$.196)
10	-5.115** ($\pm$.415)	-4.877** ($\pm$.488)	-4.756** ($\pm$.449)	-3.851** ($\pm$.494)	-2.175** ($\pm$.402)	-2.023** ($\pm$.299)	-1.474** ($\pm$.272)	-.797* ($\pm$.199)	-.432 ($\pm$.190)		.134 ($\pm$.095)
11	-5.249** ($\pm$.429)	-5.011** ($\pm$.500)	-4.890** ($\pm$.466)	-3.985** ($\pm$.507)	-2.309** ($\pm$.425)	-2.157** ($\pm$.296)	-1.608** ($\pm$.276)	-.931** ($\pm$.196)	-.566 ($\pm$.196)	-.134 ($\pm$.095)	

**P<.01 * P<.05

Pairwise comparison table depicting results based on the results of a one-way ANOVA depicting the relationship between Slope Error and the nSweeps condition for the hidden component of the newborn data at five iterations.

SE	1	2	3	4	5	6	7	8	9	10	11
1		-24.427 (±11.002)	-37.256 (±10.887)	-43.947 (±11.862)	-81.644** (±11.004)	-87.299** (±11.299)	- 104.573** (±9.460)	- 104.539** (±11.277)	- 114.864** (±10.293)	- 115.117** (±11.597)	- 121.730** (±10.655)
2	24.427 (±11.002)		-12.829 (±10.717)	-19.520 (±10.175)	-57.218** (±9.962)	-62.872** (±12.291)	-80.146** (±9.845)	-80.112** (±10.840)	-90.437** (±10.546)	-90.690** (±11.940)	-97.303** (±10.930)
3	37.256 (±10.887)	12.829 (±10.717)		-6.691 (±12.365)	-44.389** (±9.799)	-50.043** (±11.443)	-67.317** (±9.671)	-67.283** (±11.560)	-77.608** (±10.834)	-77.861** (±11.855)	-84.474** (±10.806)
4	43.947 (±11.862)	19.520 (±10.175)	6.691 (±12.365)		-37.698** (±8.128)	-43.352* (±10.698)	-60.626** (±8.287)	-60.592** (±9.554)	-70.917** (±8.068)	-71.170** (±8.983)	-77.783** (±8.432)
5	81.644** (±11.004)	57.218** (±9.962)	44.389** (±9.799)	37.698** (±8.128)		-5.654 (±9.665)	-22.929 (±6.892)	-22.895 (±8.285)	-33.219** (±7.452)	-33.472* (±8.085)	-40.086** (±7.961)
6	87.299** (±11.299)	62.872** (±12.291)	50.043** (±11.443)	43.352* (±10.698)	5.654 (±9.665)		-17.274 (±6.967)	-17.240 (±9.015)	-27.565 (±7.471)	-27.818 (±8.449)	-34.432* (±8.002)
7	104.573** (±9.460)	80.146** (±9.845)	67.317** (±9.671)	60.626** (±8.287)	22.929 (±6.892)	17.274 (±6.967)		.034 (±5.332)	-10.291 (±5.392)	-10.544 (±6.254)	-17.157 (±5.315)
8	104.539** (±11.277)	80.112** (±10.840)	67.283** (±11.560)	60.592** (±9.554)	22.895 (±8.285)	17.240 (±9.015)	-.034 (±5.332)		-10.32 (±4.703)	-10.578 (±3.633)	-17.191** (±3.698)
9	114.864** (±10.293)	90.437** (±10.546)	77.608** (±10.834)	70.917** (±8.068)	33.219** (±7.452)	27.565 (±7.471)	10.291 (±5.392)	10.325 (±4.703)		-.253 (±3.272)	-6.866 (±2.728)
10	115.117** (±11.597)	90.690** (±11.940)	77.861** (±11.855)	71.170** (±8.983)	33.472* (±8.085)	27.818 (±8.449)	10.544 (±6.254)	10.578 (±3.633)	.253 (±3.272)		-6.613 (±2.469)
11	121.730** (±10.655)	97.303** (±10.930)	84.474** (±10.806)	77.783** (±8.432)	40.086** (±7.961)	34.432* (±8.002)	17.157 (±5.315)	17.191** (±3.698)	6.866 (±2.728)	6.613 (±2.469)	

**P<.01 * P<.05

Pairwise comparison table depicting results based on the results of a one-way ANOVA depicting the relationship between Tracking Accuracy and the nSweeps condition for the hidden component of the newborn data at five iterations.

TA	1	2	3	4	5	6	7	8	9	10	11
1		-.124 (±.038)	-.108 (±.039)	-.122 (±.046)	-.158** (±.033)	-.153* (±.038)	-.209** (±.037)	-.242** (±.037)	-.257** (±.040)	-.277** (±.039)	-.298** (±.037)
2	.124 (±.038)		.016 (±.045)	.003 (±.048)	-.034 (±.046)	-.029 (±.052)	-.085 (±.042)	-.118 (±.040)	-.133 (±.043)	-.153* (±.039)	-.174** (±.038)
3	.108 (±.039)	-.016 (±.045)		-.014 (±.039)	-.050 (±.050)	-.045 (±.050)	-.101 (±.038)	-.134 (±.036)	-.149* (±.037)	-.169** (±.038)	-.190** (±.035)
4	.122 (±.046)	-.003 (±.048)	.014 (±.039)		-.036 (±.047)	-.031 (±.055)	-.087 (±.045)	-.121 (±.040)	-.135 (±.042)	-.156 (±.043)	-.177* (±.042)
5	.158** (±.033)	.034 (±.046)	.050 (±.050)	.036 (±.047)		.005 (±.036)	-.051 (±.037)	-.084 (±.036)	-.099 (±.040)	-.119 (±.039)	-.140 (±.039)
6	.153* (±.038)	.029 (±.052)	.045 (±.050)	.031 (±.055)	-.005 (±.036)		-.056 (±.027)	-.089 (±.030)	-.104 (±.034)	-.124 (±.036)	-.146* (±.034)
7	.209** (±.037)	.085 (±.042)	.101 (±.038)	.087 (±.045)	.051 (±.037)	.056 (±.027)		-.033 (±.016)	-.048 (±.021)	-.068 (±.022)	-.089** (±.020)
8	.242** (±.037)	.118 (±.040)	.134 (±.036)	.121 (±.040)	.084 (±.036)	.089 (±.030)	.033 (±.016)		-.015 (±.019)	-.035 (±.019)	-.056 (±.016)
9	.257** (±.040)	.133 (±.043)	.149* (±.037)	.135 (±.042)	.099 (±.040)	.104 (±.034)	.048 (±.021)	.015 (±.019)		-.020 (±.011)	-.041 (±.012)
10	.277** (±.039)	.153* (±.039)	.169** (±.038)	.156 (±.043)	.119 (±.039)	.124 (±.036)	.068 (±.022)	.035 (±.019)	.020 (±.011)		-.021 (±.008)
11	.298** (±.037)	.174* (±.038)	.190** (±.035)	.177* (±.042)	.140 (±.039)	.146* (±.034)	.089** (±.020)	.056 (±.016)	.041 (±.012)	.021 (±.008)	

**P<.01 * P<.05

Pairwise comparison table depicting results based on the results of a two-way ANOVA depicting the relationship between Frequency Error and the nSweeps condition for the newborn and adult data at five iterations.

FE	1	2	3	4	5	6	7	8	9	10	11
1		-.553 (±.300)	-1.891** (±.336)	-3.321** (±.356)	-4.842** (±.396)	-5.607** (±.347)	-6.219** (±.329)	-6.407** (±.351)	-6.778** (±.334)	-6.985** (±.329)	-7.138** (±.353)
2	.553 (±.300)		-1.338** (±.232)	-2.768** (±.326)	-4.289** (±.351)	-5.054** (±.296)	-5.666** (±.283)	-5.853** (±.312)	-6.224** (±.291)	-6.432** (±.298)	-6.585** (±.328)
3	1.891** (±.336)	1.338** (±.232)		-1.430** (±.276)	-2.951** (±.308)	-3.716** (±.291)	-4.328** (±.267)	-4.516** (±.290)	-4.887** (±.300)	-5.094** (±.303)	-5.247** (±.327)
4	3.321** (±.356)	2.768** (±.326)	1.430** (±.276)		-1.521** (±.246)	-2.286** (±.269)	-2.898** (±.237)	-3.086** (±.250)	-3.456** (±.267)	-3.664** (±.281)	-3.817** (±.299)
5	4.842** (±.396)	4.289** (±.351)	2.951** (±.308)	1.521** (±.246)		-.765 (±.240)	-1.377** (±.198)	-1.565** (±.237)	-1.935** (±.238)	-2.143** (±.264)	-2.296** (±.267)
6	5.607** (±.347)	5.054** (±.296)	3.716** (±.291)	2.286** (±.269)	.765 (±.240)		-.612 (±.178)	-.800** (±.199)	-1.170** (±.187)	-1.378** (±.215)	-1.531** (±.235)
7	6.219** (±.329)	5.666** (±.283)	4.328** (±.267)	2.898 (±.237)	1.377** (±.198)	.612 (±.178)		-.188 (±.156)	-.559** (±.136)	-.766** (±.143)	-.919** (±.142)
8	6.407** (±.351)	5.853** (±.312)	4.516** (±.290)	3.086** (±.250)	1.565** (±.237)	.800** (±.199)	.188 (±.156)		-.371 (±.172)	-.578* (±.165)	-.731* (±.202)
9	6.778** (±.334)	6.224** (±.291)	4.887** (±.300)	3.456** (±.267)	1.935** (±.238)	1.170** (±.187)	.559** (±.136)	.371 (±.172)		-.208 (±.090)	-.360 (±.127)
10	6.985** (±.329)	6.432** (±.298)	5.094** (±.303)	3.664** (±.281)	2.143** (±.264)	1.378** (±.215)	.766** (±.143)	.578* (±.165)	.208 (±.090)		-.153 (±.093)
11	7.138** (±.353)	6.585** (±.328)	5.247** (±.327)	3.817** (±.299)	2.296** (±.267)	1.531** (±.235)	.919** (±.142)	.731* (±.202)	.360 (±.127)	.153 (±.093)	

**P<.01 * P<.05

Pairwise comparison table depicting results based on the results of a two-way ANOVA depicting the relationship between Slope Error and the nSweeps condition for the newborn and adult data at five iterations.

SE	1	2	3	4	5	6	7	8	9	10	11
1		19.748 (±8.258)	37.049** (±8.319)	60.020** (±8.821)	78.090** (±9.305)	85.184** (±8.115)	97.019** (±8.314)	97.437** (±8.714)	105.014** (±8.301)	104.433** (±8.648)	109.333** (±8.896)
2	-19.748 (±8.258)		17.301 (±6.854)	40.271** (±7.759)	58.342** (±7.674)	65.435** (±7.581)	77.271** (±6.837)	77.688** (±8.149)	85.266** (±7.611)	84.685** (±8.448)	89.584** (±8.266)
3	-37.049** (±8.319)	-17.301 (±6.854)		22.970 (±6.954)	41.041** (±7.061)	48.135** (±6.797)	59.970** (±6.532)	60.387** (±7.748)	67.965** (±7.599)	67.384** (±7.905)	72.283** (±8.028)
4	-60.020** (±8.821)	-40.271** (±7.759)	-22.970 (±6.954)		18.070 (±5.523)	25.164** (±5.399)	37.000** (±5.839)	37.417** (±6.593)	44.995** (±6.419)	44.413** (±6.886)	49.313** (±6.659)
5	-78.090** (±9.305)	-58.342** (±7.674)	-41.041** (±7.061)	-18.070 (±5.523)		7.094 (±4.140)	18.929* (±4.963)	19.347 (±6.180)	26.924** (±6.171)	26.343** (±6.430)	31.243** (±6.361)
6	-85.184** (±8.115)	-65.435** (±7.581)	-48.135** (±6.797)	-25.164** (±5.399	-7.094 (±4.140)		11.835 (±4.505)	12.253 (±4.728)	19.831** (±4.826)	19.249** (±4.713)	24.149** (±5.220)
7	-97.019** (±8.314)	-77.271** (±6.837)	-59.970** (±6.532)	-37.000** (±5.839)	-18.929* (±4.963)	-11.835 (±4.505)		.417 (±3.117)	7.995 (±3.023)	7.414 (±3.506)	12.314** (±3.079)
8	-97.437** (±8.714)	-77.688** (±8.149)	-60.387** (±7.748)	-37.417** (±6.593)	-19.347 (±6.180)	-12.253 (±4.728)	-.417 (±3.117)		7.578 (±3.060)	6.997 (±3.035)	11.896** (±3.250)
9	-105.014** (±8.301)	-85.266** (±7.611)	-67.965** (±7.599)	-44.995** (±6.419)	-26.924** (±6.171)	-19.831** (±4.826)	-7.995 (±3.023)	-7.578 (±3.060)		-.581 (±2.068)	4.318 (±1.972)
10	-104.433** (±8.648)	-84.685** (±8.448)	-67.384** (±7.905)	-44.413** (±6.886)	-26.343** (±6.430)	-19.249** (±4.713)	-7.414 (±3.506)	-6.997 (±3.035)	.581 (±2.068)		4.900 (±1.941)
11	-109.333** (±8.896)	-89.584** (±8.266)	-72.283** (±8.028)	-49.313** (±6.659)	-31.243** (±6.361)	-24.149** (±5.220)	-12.314** (±3.079)	-11.896* (±3.250)	-4.318 (±1.972)	-4.900 (±1.941)	

**P<.01 * P<.05

Pairwise comparison table depicting results based on the results of a two-way ANOVA depicting the relationship between Tracking Accuracy and the nSweeps condition for the newborn and adult data at five iterations.

TA	1	2	3	4	5	6	7	8	9	10	11
1		.034 (±.018)	.103** (±.020)	.190** (±.018)	.261** (±.021)	.292** (±.021)	.327** (±.023)	.344** (±.021)	.365** (±.020)	.377** (±.021)	.388** (±.022)
2	-.034 (±.018)		.070** (±.016)	.156** (±.018)	.227** (±.019)	.258** (±.020)	.293** (±.020)	.310** (±.019)	.331** (±.019)	.343** (±.019)	.354** (±.020)
3	-.103** (±.020)	-.070** (±.016)		.087** (±.017)	.158** (±.017)	.189** (±.019)	.223** (±.020)	.241** (±.020)	.261** (±.020)	.274** (±.020)	.285** (±.022)
4	-.190** (±.018)	-.156** (±.018)	-.087** (±.017)		.071** (±.014)	.102** (±.016)	.136** (±.019)	.154** (±.017)	.174** (±.017)	.187** (±.018)	.198** (±.020)
5	-.261** (±.021)	-.227** (±.01)	-.158** (±.017)	-.071** (±.014)		.031 (±.013)	.066** (±.014)	.083** (±.014)	.104** (±.015)	.116** (±.017)	.127** (±.017)
6	-.292** (±.021)	-.258** (±.020)	-.189** (±.019)	-.102** (±.016)	-.031 (±.013)		.034 (±.015)	.052** (±.013)	.072** (±.013)	.085** (±.015)	.096** (±.016)
7	-.327** (±.023)	-.293** (±.020)	-.223** (±.020)	-.136** (±.019)	-.066** (±.014)	-.034 (±.015)		.018 (±.010)	.038* (±.010)	.050** (±.012)	.061** (±.010)
8	-.344** (±.021)	-.310** (±.019)	-.241** (±.020)	-.154** (±.017)	-.083** (±.014)	-.052** (±.013)	-.018 (±.010)		.020 (±.011)	.033 (±.012)	.044* (±.012)
9	-.365 (±.020)	-.331** (±.019)	-.261** (±.020)	-.174** (±.017)	-.104** (±.015)	-.072** (±.013)	-.038* (±.010)	-.020 (±.011)		.012 (±.008)	.023 (±.009)
10	-.377** (±.021)	-.343** (±.019)	-.274** (±.020)	-.187** (±.018)	-.116** (±.017)	-.085** (±.015)	-.050** (±.012)	-.033 (±.012)	-.012 (±.008)		.011 (±.008)
11	-.388** (±.022)	-.354** (±.020)	-.285** (±.022)	-.198** (±.020)	-.127** (±.017)	-.096** (±.016)	-.061** (±.010)	-.044* (±.012)	-.023 (±.009)	-.011 (±.008)	

**P<.01 * P<.05

Pairwise comparison table depicting results based on the results of a two-way ANOVA depicting the relationship between Frequency Error and the nSweeps condition for the adult data at five iterations.

FE	1	2	3	4	5	6	7	8	9	10	11
1		-1.814** (±.411)	-3.632** (±.441)	-5.147** (±.480)	-6.485** (±.477)	-6.889** (±.478)	-7.226** (±.488)	-7.455** (±.497)	-7.563** (±.507)	-7.738** (±.495)	-7.792** (±.507)
2	1.814** (±.411)		-1.818** (±.355)	-3.333** (±.432)	-4.671** (±.430)	-5.074** (±.433)	-5.411** (±.430)	-5.641** (±.481)	-5.748** (±.476)	-5.923** (±.471)	-5.978** (±.492)
3	3.632** (±.441)	1.818** (±.355)		-1.515** (±.354)	-2.852** (±.354)	-3.256** (±.418)	-3.593** (±.396)	-3.823** (±.461)	-3.930** (±.454)	-4.105** (±.446)	-4.159** (±.465)
4	5.147** (±.480)	3.333** (±.432)	1.515** (±.354)		-1.337** (±.270)	-1.741** (±.336)	-2.078** (±.335)	-2.308** (±.326	-2.415** (±.341)	-2.590** (±.384)	-2.644** (±.425)
5	6.485** (±.477)	4.671** (±.430)	2.852** (±.354)	1.337** (±.270)		-.404 (±.229)	-.741** (±.170)	-.970** (±.204)	-1.078** (±.185)	-1.253** (±.226)	-1.307** (±.210)
6	6.889** (±.478)	5.074** (±.433)	3.256** (±.418)	1.741** (±.336)	.404 (±.229)		-.337 (±.204)	-.566 (±.215)	-.674 (±.230)	-.849 (±.256)	-.903 (±.261)
7	7.226** (±.488)	5.411** (±.430)	3.593** (±.396)	2.078** (±.335)	.741** (±.170)	.337 (±.204)		-.229 (±.151)	-.337 (±.167)	-.512 (±.165)	-.566 (±.158)
8	7.455** (±.497)	5.641** (±.481)	3.823** (±.461)	2.308** (±.326)	.970** (±.204)	.566 (±.215)	.229 (±.151)		-.108 (±.129)	-.283 (±.172)	-.337 (±.202)
9	7.563** (±.507)	5.748** (±.476)	3.930** (±.454)	2.415** (±.341)	1.078** (±.185)	.674 (±.230)	.337 (±.167)	.108 (±.129)		-.175 (±.102)	-.229 (±.152)
10	7.738** (±.495)	5.923** (±.471)	4.105** (±.446)	2.590** (±.384)	1.253** (±.226)	.849 (±.256)	.512 (±.165)	.283 (±.172)	.175 (±.102)		-.054 (±.119)
11	7.792** (±.507)	5.978** (±.492)	4.159** (±.465)	2.644** (±.425)	1.307** (±.210)	.903 (±.261)	.566 (±.158)	.337 (±.202)	.229 (±.152)	.054 (±.119)	

**P<.01 * P<.05

Pairwise comparison table depicting results based on the results of a two-way ANOVA depicting the relationship between Slope Error and the nSweeps condition for the adult data at five iterations.

SE	1	2	3	4	5	6	7	8	9	10	11
1		46.078 (±12.893)	69.466** (±11.310)	96.486** (±11.543)	109.83** (±12.805)	109.802** (±12.036)	111.811** (±12.949)	112.279** (±13.566)	116.007** (±13.655)	116.118** (±14.162)	118.530** (±14.398)
2	-46.078 (±12.893)		23.388 (±9.148)	50.409** (±7.985)	63.754** (±9.098)	63.724** (±10.627)	65.733** (±10.658)	66.201** (±12.192)	69.930** (±12.025)	70.041** (±13.260)	72.452** (±13.103)
3	-69.466** (±11.310)	-23.388 (±9.148)		27.020* (±6.980)	40.366** (±7.489)	40.336** (±8.894)	42.345** (±8.432)	42.813* (±11.038)	46.541* (±11.059)	46.653* (±11.926)	49.064* (±11.775)
4	-96.486** (±11.543)	-50.409** (±7.985)	-27.020 (±6.980)		13.345 (±5.287)	13.316 (±6.445)	15.325 (±6.749)	15.793 (±9.289)	19.521 (±8.621)	19.632 (±9.858)	22.043 (±9.739)
5	-109.832** (±12.805)	-63.754** (±9.098)	-40.366** (±7.489)	-13.345 (±5.287)		-.029 (±5.070)	1.979 (±4.909)	2.447 (±7.473)	6.176 (±6.921)	6.287 (±7.813)	8.698 (±7.533)
6	-109.802** (±12.036)	-63.724** (±10.627)	-40.336** (±8.894)	-13.316 (±6.445)	.029 (±5.070)		2.009 (±4.388)	2.477 (±5.657)	6.205 (±5.579)	6.316 (±6.162)	8.728 (±6.486)
7	-111.811** (±12.949)	-65.733** (±10.658)	-42.345** (±8.432)	-15.325 (±6.749)	-1.979 (±4.909)	-2.009 (±4.388)		.468 (±4.182)	4.196 (±3.979)	4.308 (±4.869)	6.719 (±4.869)
8	-112.279** (±13.566)	-66.201** (±12.192)	-42.813* (±11.038)	-15.793 (±9.289)	-2.447 (±7.473)	-2.477 (±5.657)	-.468 (±4.182)		3.728 (±3.858)	3.840 (±4.108)	6.251 (±5.088)
9	-116.007** (±13.655)	-69.930** (±12.025)	-46.541* (±11.059)	-19.521 (±8.621)	-6.176 (±6.921)	-6.205 (±5.579)	-4.196 (±3.979)	-3.728 (±3.858)		.111 (±1.720)	2.522 (±1.830)
10	-116.118** (±14.162)	-70.041** (±13.260)	-46.653* (±11.926)	-19.632 (±9.858)	-6.287 (±7.813)	-6.316 (±6.162)	-4.308 (±4.869)	-3.840 (±4.108)	-.111 (±1.720)		2.411 (±1.540)
11	-118.530** (±14.398)	-72.452** (±13.103)	-49.064* (±11.775)	-22.043 (±9.739)	-8.698 (±7.533)	-8.728 (±6.486)	-6.719 (±4.869)	-6.251 (±5.088)	-2.522 (±1.830)	-2.411 (±1.540)	

**P<.01 * P<.05

Pairwise comparison table depicting results based on the results of a two-way ANOVA depicting the relationship between Tracking Accuracy and the nSweeps condition for the adult data at five iterations.

TA	1	2	3	4	5	6	7	8	9	10	11
1		.086* (±.022)	.190** (±.023)	.270** (±.022)	.329** (±.025)	.346** (±.027)	.355** (±.029)	.371** (±.029)	.375** (±.028)	.390** (±.029)	.384** (±.033)
2	-.086* (±.022)		.104* (±.025)	.183** (±.023)	.243** (±.024)	.260** (±.026)	.269** (±.028)	.284** (±.028)	.289** (±.027)	.303** (±.028)	.298** (±.031)
3	-.190** (±.023)	-.104* (±.025)		.080** (±.019)	.139** (±.021)	.156** (±.026)	.165** (±.027)	.181** (±.029)	.185** (±.028)	.200** (±.026)	.194** (±.032)
4	-.270** (±.022)	-.183** (±.023)	-.080* (±.019)		.060 (±.017)	.076 (±.022)	.085* (±.022)	.101** (±.022)	.106** (±.021)	.120** (±.022)	.115* (±.028)
5	-.329** (±.025)	-.243** (±.024)	-.139** (±.021)	-.060 (±.017)		.017 (±.014)	.026 (±.010)	.041** (±.014)	.046* (±.011)	.061** (±.013)	.055 (±.016)
6	-.346** (±.027)	-.260** (±.026)	-.156** (±.026)	-.076 (±.022)	-.017 (±.014)		.009 (±.014)	.025 (±.014)	.029 (±.015)	.044 (±.015)	.038 (±.018)
7	-.355** (±.029)	-.269** (±.028)	-.165** (±.027)	-.085* (±.022)	-.026 (±.010)	-.009 (±.014)		.016 (±.010)	.020 (±.009)	.035 (±.012)	.029 (±.011)
8	-.371** (±.029)	-.284** (±.028)	-.181** (±.029)	-.101** (±.022)	-.041 (±.014)	-.025 (±.014)	-.016 (±.010)		.005 (±.012)	.019 (±.014)	.014 (±.017)
9	-.375** (±.028)	-.289** (±.027)	-.185** (±.028)	-.106** (±.021)	-.046** (±.011)	-.029 (±.015)	-.020 (±.009)	-.005 (±.012)		.015 (±.009)	.009 (±.010)
10	-.390** (±.029)	-.303** (±.028)	-.200** (±.026)	-.120** (±.022)	-.061** (±.013)	-.044 (±.015)	-.035 (±.012)	-.019 (±.014)	-.015 (±.009)		-.005 (±.013)
11	-.384** (±.033)	-.298 ** (±.031)	-.194** (±.032)	-.115* (±.028)	-.055 (±.016)	-.038 (±.018)	-.029 (±.011)	-.014 (±.017)	-.009 (±.010)	.005 (±.013)	

**P<.01 * P<.05

Pairwise comparison table depicting results based on the results of a two-way ANOVA depicting the relationship between Frequency Error and the nSweeps condition for the newborn data at five iterations.

FE	1	2	3	4	5	6	7	8	9	10	11
1		.707 (±.436)	-.150 (±.508)	-1.495 (±.526)	-3.199** (±.632)	-4.326** (±.503)	-5.212** (±.442)	-5.359** (±.495)	-5.992** (±.436)	-6.233** (±.433)	-6.484** (±.491)
2	-.707 (±.436)		-.857 (±.299)	-2.203** (±.488)	-3.907** (±.556)	-5.033** (±.403)	-5.920** (±.367)	-6.066** (±.397)	-6.700** (±.337)	-6.940** (±.365)	-7.192** (±.434)
3	.150 (±.508)	.857 (±.299)		-1.346 (±.424)	-3.050** (±.504)	-4.176** (±.405)	-5.063** (±.359)	-5.209** (±.353)	-5.843** (±.392)	-6.083** (±.408)	-6.335** (±.461)
4	1.495 (±.526)	2.203** (±.488)	1.346 (±.424)		-1.704* (±.410)	-2.831** (±.421)	-3.717** (±.337)	-3.863** (±.378)	-4.497** (±.411)	-4.737** (±.412)	-4.989** (±.422)
5	3.199** (±.632)	3.907** (±.556)	3.050** (±.504)	1.704* (±.410)		-1.127 (±.422)	-2.013** (±.357)	-2.159** (±.427)	-2.793** (±.439)	-3.033** (±.477)	-3.285** (±.492)
6	4.326** (±.503)	5.033** (±.403)	4.176** (±.405)	2.831** (±.421)	1.127 (±.422)		-.886 (±.292)	-1.033 (±.335)	-1.667** (±.295)	-1.907** (±.346)	-2.158** (±.391)
7	5.212** (±.442)	5.920** (±.367)	5.063** (±.359)	3.717** (±.337)	2.013** (±.357)	.886 (±.292)		-.146 (±.273)	-.780 (±.214)	-1.020** (±.235)	-1.272** (±.236)
8	5.359** (±.495)	6.066** (±.397)	5.209** (±.353)	3.863** (±.378)	2.159** (±.427)	1.033 (±.335)	.146 (±.273)		-.634 (±.319)	-.874 (±.280)	-1.126 (±.349)
9	5.992** (±.436)	6.700** (±.337)	5.843** (±.392)	4.497** (±.411)	2.793** (±.439)	1.667** (±.295)	.780 (±.214)	.634 (±.319)		-.240 (±.148)	-.492 (±.203)
10	6.233** (±.433)	6.940** (±.365)	6.083** (±.408)	4.737** (±.412)	3.033** (±.477)	1.907** (±.346)	1.020** (±.235)	.874 (±.280)	.240 (±.148)		-.251 (±.143)
11	6.484** (±.491)	7.192** (±.434)	6.335** (±.461)	4.989** (±.422)	3.285** (±.492)	2.158** (±.391)	1.272** (±.236)	1.126 (±.349)	.492 (±.203)	.251 (±.143)	

**P<.01 * P<.05

Pairwise comparison table depicting results based on the results of a two-way ANOVA depicting the relationship between Slope Error and the nSweeps condition for the newborn data at five iterations.

SE	1	2	3	4	5	6	7	8	9	10	11
1		-6.581 ($\pm$10.323)	4.633 ($\pm$12.204)	23.553 ($\pm$13.342)	46.348 ($\pm$13.505)	60.565** ($\pm$10.888)	82.228** ($\pm$10.432)	82.594** ($\pm$10.942)	94.022** ($\pm$9.444)	92.748** ($\pm$9.928)	100.136** ($\pm$10.454)
2	6.581 ($\pm$10.323)		11.214 ($\pm$10.210)	30.134 ($\pm$13.305)	52.930** ($\pm$12.362)	67.146** ($\pm$10.815)	88.809** ($\pm$8.567)	89.175** ($\pm$10.814)	100.603** ($\pm$9.333)	99.329** ($\pm$10.471)	106.717** ($\pm$10.079)
3	-4.633 ($\pm$12.204)	-11.214 ($\pm$10.210)		18.920 ($\pm$12.031)	41.716 ($\pm$11.972)	55.933** ($\pm$10.281)	77.595** ($\pm$9.979)	77.962** ($\pm$10.875)	89.389** ($\pm$10.426)	88.115** ($\pm$10.379)	95.503** ($\pm$10.916)
4	-23.553 ($\pm$13.342)	-30.134 ($\pm$13.305)	-18.920 ($\pm$12.031)		22.795 ($\pm$9.699)	37.012** ($\pm$8.665)	58.675** ($\pm$9.530)	59.041** ($\pm$9.357)	70.469** ($\pm$9.514)	69.195** ($\pm$9.617)	76.583** ($\pm$9.085)
5	-46.348 ($\pm$13.505)	-52.930** ($\pm$12.362)	-41.716 ($\pm$11.972)	-22.795 ($\pm$9.699)		14.217 ($\pm$6.546)	35.879* ($\pm$8.628)	36.246 ($\pm$9.846)	47.673** ($\pm$10.219)	46.399** ($\pm$10.213)	53.787** ($\pm$10.251)
6	-60.565** ($\pm$10.888)	-67.146** ($\pm$10.815)	-55.933** ($\pm$10.281)	-37.012** ($\pm$8.665)	-14.217 ($\pm$6.546)		21.662 ($\pm$7.868)	22.029 ($\pm$7.577)	33.456* ($\pm$7.876)	32.182** ($\pm$7.133)	39.570** ($\pm$8.182)
7	-82.228** ($\pm$10.432)	-88.809** ($\pm$8.567)	-77.595** ($\pm$9.979)	-58.675** ($\pm$9.530)	-35.879* ($\pm$8.628)	-21.662 ($\pm$7.868)		.367 ($\pm$4.622)	11.794 ($\pm$4.553)	10.520 ($\pm$5.046)	17.908** ($\pm$3.772)
8	-82.594** ($\pm$10.942)	-89.175** ($\pm$10.814)	-77.962** ($\pm$10.875)	-59.041** ($\pm$9.357)	-36.246 ($\pm$9.846)	-22.029 ($\pm$7.577)	-.367 ($\pm$4.622)		11.427 ($\pm$4.750)	10.154 ($\pm$4.468)	17.542** ($\pm$4.046)
9	-94.022** ($\pm$9.444)	- 100.603** ($\pm$9.333)	-89.389** ($\pm$10.426)	-70.469** ($\pm$9.514)	-47.673** ($\pm$10.219)	-33.456* ($\pm$7.876)	-11.794 ($\pm$4.553)	-11.427 ($\pm$4.750)		-1.274 ($\pm$3.762)	6.114 ($\pm$3.494)
10	-92.748** ($\pm$9.928)	-99.329** ($\pm$10.471)	-88.115** ($\pm$10.379)	-69.195** ($\pm$9.617)	-46.399** ($\pm$10.213)	-32.182** ($\pm$7.133)	-10.520 ($\pm$5.046)	-10.154 ($\pm$4.468)	1.274 ($\pm$3.762)		7.388 ($\pm$3.564)
11	- 100.136** ($\pm$10.454)	- 106.717** ($\pm$10.079)	-95.503** ($\pm$10.916)	-76.583** ($\pm$9.085)	-53.787** ($\pm$10.251)	-39.570** ($\pm$8.182)	-17.908 ($\pm$3.772)	-17.542** ($\pm$4.046)	-6.114 ($\pm$3.494)	-7.388 ($\pm$3.564)	

**P<.01 * P<.05

Pairwise comparison table depicting results based on the results of a two-way ANOVA depicting the relationship between Tracking Accuracy and the nSweeps condition for the newborn data at five iterations.

TA	1	2	3	4	5	6	7	8	9	10	11
1		-.019 (±.029)	.017 (±.032)	.111* (±.029)	.193** (±.034)	.239** (±.032)	.298** (±.036)	.318** (±.031)	.354** (±.029)	.364** (±.030)	.392** (±.029)
2	.019 (±.029)		.036 (±.020)	.130** (±.029)	.212** (±.031)	.257** (±.030)	.317** (±.028)	.337** (±.025)	.373** (±.025)	.383** (±.026)	.410** (±.025)
3	-.017 (±.032)	-.036 (±.020)		.094 (±.028)	.176** (±.027)	.222** (±.028)	.282** (±.030)	.301** (±.027)	.337** (±.029)	.347** (±.029)	.375** (±.030)
4	-.111* (±.029)	-.130** (±.029)	-.094 (±.028)		.082 (±.023)	.128** (±.025)	.187** (±.030)	.207** (±.026)	.243** (±.027)	.253** (±.029)	.281** (±.027)
5	-.193** (±.034)	-.212** (±.031)	-.176** (±.027)	-.082 (±.023)		.046 (±.022)	.105* (±.025)	.125 (±.025)	.161** (±.027)	.171** (±.032)	.199** (±.030)
6	-.239** (±.032)	-.257** (±.030)	-.222** (±.028)	-.128** (±.025)	-.046 (±.022)		.060 (±.026)	.079 (±.022)	.115** (±.022)	.125** (±.026)	.153** (±.026)
7	-.298** (±.036)	-.317** (±.028)	-.282** (±.030)	-.187** (±.030)	-.105* (±.025)	-.060 (±.026)		.020 (±.018)	.056 (±.019)	.066 (±.021)	.093** (±.017)
8	-.318** (±.031)	-.337** (±.025)	-.301** (±.027)	-.207** (±.026)	-.125** (±.025)	-.079 (±.022)	-.020 (±.018)		.036 (±.019)	.046 (±.019)	.074* (±.018)
9	-.354** (±.029)	-.373** (±.025)	-.337** (±.029)	-.243** (±.027)	-.161** (±.027)	-.115** (±.022)	-.056 (±.019)	-.036 (±.019)		.010 (±.014)	.038 (±.014)
10	-.364** (±.030)	-.383** (±.026)	-.347** (±.029)	-.253** (±.029)	-.171** (±.032)	-.125** (±.026)	-.066 (±.021)	-.046 (±.019)	-.010 (±.014)		.028 (±.010)
11	-.392** (±.029)	-.410** (±.025)	-.375** (±.030)	-.281** (±.027)	-.199** (±.030)	-.153** (±.026)	-.093** (±.017)	-.074* (±.018)	-.038 (±.014)	-.028 (±.010)	

**P<.01 * P<.05

Appendix E

10 Iteration nSweeps Confusion Matrices

Pairwise comparison table depicting results based on the results of a one-way ANOVA depicting the relationship between Frequency Error and the nSweeps condition for the hidden component of newborn and adult data at ten iterations.

FE	1	2	3	4	5	6	7	8	9	10	11
1		.362 (±.285)	.595 (±.346)	1.696** (±.370)	2.926** (±.365)	3.441** (±.387)	3.941** (±.401)	4.339** (±.375)	4.719** (±.356)	4.785** (±.370)	4.923** (±.364)
2	-.362 (±.285)		.234 (±.297)	1.334** (±.328)	2.564** (±.337)	3.079** (±.366)	3.580** (±.397)	3.977** (±.376)	4.357** (±.358)	4.423** (±.371)	4.561** (±.363)
3	-.595 (±.346)	-.234 (±.297)		1.101** (±.233)	2.331** (±.299)	2.845** (±.321)	3.346** (±.337)	3.743** (±.322)	4.124** (±.317)	4.190** (±.328)	4.328** (±.327)
4	-1.696** (±.370)	-1.334** (±.328)	-1.101** (±.233)		1.230** (±.171)	1.745** (±.210)	2.245** (±.228)	2.643** (±.211)	3.023** (±.205)	3.089** (±.215)	3.227** (±.227)
5	-2.926** (±.365)	-2.564** (±.337)	-2.331** (±.299)	-1.230** (±.171)		.515* (±.137)	1.015** (±.170)	1.413** (±.156)	1.793** (±.166)	1.859** (±.166)	1.997** (±.182)
6	-3.441** (±.387)	-3.079** (±.366)	-2.845** (±.321)	-1.745** (±.210)	-.515* (±.137)		.501* (±.126)	.898** (±.132)	1.278** (±.143)	1.344** (±.154)	1.482** (±.166)
7	-3.941** (±.401)	-3.580** (±.397)	-3.346** (±.337)	-2.245** (±.228)	-1.015** (±.170)	-.501* (±.126)		.397* (±.108)	.778** (±.115)	.844** (±.103)	.982** (±.122)
8	-4.339** (±.375)	-3.977** (±.376)	-3.743** (±.322)	-2.643** (±.211)	-1.413** (±.156)	-.898** (±.132)	-.397* (±.108)		.380** (±.091)	.446** (±.094)	.584** (±.111)
9	-4.719** (±.356)	-4.357** (±.358)	-4.124** (±.317)	-3.023** (±.205)	-1.793** (±.166)	-1.278** (±.143)	-.778** (±.115)	-.380** (±.091)		.066 (±.046)	.204* (±.057)
10	-4.785** (±.370)	-4.423** (±.371)	-4.190** (±.328)	-3.089** (±.215)	-1.859** (±.166)	-1.344** (±.154)	-.844** (±.103)	-.446** (±.094)	-.066 (±.046)		.138 (±.047)
11	-4.923** (±.364)	-4.561** (±.363)	-4.328** (±.327)	-3.227** (±.227)	-1.997** (±.182)	-1.482** (±.166)	-.982** (±.122)	-.584** (±.111)	-.204* (±.057)	-.138 (±.047)	

**P<.01 * P<.05

Pairwise comparison table depicting results based on the results of a one-way ANOVA depicting the relationship between Slope Error and the nSweeps condition for the hidden component of newborn and adult data at ten iterations.

SE	1	2	3	4	5	6	7	8	9	10	11
1		-15.096 (±5.160)	-32.249** (±6.227)	-56.218** (±5.904)	-73.961** (±6.944)	-81.995** (±7.864)	-91.844** (±7.603)	-94.610** (±7.923)	-100.412** (±8.030)	-102.145** (±8.002)	-103.442** (±7.826)
2	15.096 (±5.160)		-17.153 (±6.305)	-41.122** (±5.893)	-58.865** (±6.445)	-66.900** (±6.950)	-76.748** (±7.009)	-79.514** (±6.833)	-85.316** (±7.325)	-87.049** (±7.378)	-88.346** (±7.278)
3	32.249** (±6.227)	17.153 (±6.305)		-23.969** (±5.405)	-41.712** (±5.404)	-49.746** (±5.623)	-59.595** (±5.643)	-62.361** (±5.948)	-68.163** (±6.385)	-69.896** (±6.359)	-71.193** (±6.419)
4	56.218** (±5.904)	41.122** (±5.893)	23.969** (±5.405)		-17.743** (±4.005)	-25.778** (±5.031)	-35.626** (±4.790)	-38.393** (±4.997)	-44.194** (±5.442)	-45.928** (±5.344)	-47.224** (±5.476)
5	73.961** (±6.944)	58.865** (±6.445)	41.712** (±5.404)	17.743** (±4.005)		-8.035 (±3.347)	-17.883** (±3.429)	-20.650** (±4.022)	-26.451** (±4.555)	-28.185** (±4.552)	-29.481** (±4.820)
6	81.995** (±7.864)	66.900** (±6.950)	49.746** (±5.623)	25.778** (±5.031)	8.035 (±3.347)		-9.848* (±2.537)	-12.615** (±2.957)	-18.417** (±3.811)	-20.150** (±3.975)	-21.446** (±4.126)
7	91.844** (±7.603)	76.748** (±7.009)	59.595** (±5.643)	35.626** (±4.790)	17.883** (±3.429)	9.848* (±2.537)		-2.766 (±2.218)	-8.568* (±2.370)	-10.301** (±2.463)	-11.598** (±2.592)
8	94.610** (±7.923)	79.514** (±6.833)	62.361** (±5.948)	38.393** (±4.997)	20.650** (±4.022)	12.615** (±2.957)	2.766 (±2.218)		-5.802 (±2.231)	-7.535 (±2.538)	-8.832 (±2.782)
9	100.412** (±8.030)	85.316** (±7.325)	68.163** (±6.385)	44.194** (±5.442)	26.451** (±4.555)	18.417** (±3.811)	8.568* (±2.370)	5.802 (±2.231)		-1.733 (±1.223)	-3.030 (±1.167)
10	102.145** (±8.002)	87.049** (±7.378)	69.896** (±6.359)	45.928** (±5.344)	28.185** (±4.552)	20.150** (±3.975)	10.301** (±2.463)	7.535 (±2.538)	1.733 (±1.223)		-1.297 (±1.056)
11	103.442** (±7.826)	88.346** (±7.278)	71.193** (±6.419)	47.224** (±5.476)	29.481** (±4.820)	21.446** (±4.126)	11.598** (±2.592)	8.832 (±2.782)	3.030 (±1.167)	1.297 (±1.056)	

**P<.01 * P<.05

Pairwise comparison table depicting results based on the results of a one-way ANOVA depicting the relationship between Tracking Accuracy and the nSweeps condition for the hidden component of newborn and adult data at ten iterations.

TA	1	2	3	4	5	6	7	8	9	10	11
1		-.013 (±.021)	-.005 (±.025)	-.058 (±.022)	-.110** (±.024)	-.136** (±.026)	-.165** (±.026)	-.188** (±.025)	-.210** (±.025)	-.219** (±.025)	-.225** (±.025)
2	.013 (±.021)		.008 (±.019)	-.045 (±.020)	-.097** (±.023)	-.123** (±.024)	-.152** (±.025)	-.175** (±.023)	-.197** (±.024)	-.207** (±.024)	-.212** (±.023)
3	.005 (±.025)	-.008 (±.019)		-.053 (±.016)	-.105** (±.021)	-.131** (±.022)	-.160** (±.022)	-.183** (±.023)	-.205** (±.024)	-.215** (±.024)	-.220** (±.023)
4	.058 (±.022)	.045 (±.020)	.053 (±.016)		-.053** (±.013)	-.078** (±.015)	-.107** (±.016)	-.130** (±.015)	-.152** (±.017)	-.162** (±.017)	-.167** (±.017)
5	.110** (±.024)	.097** (±.023)	.105** (±.021)	.053** (±.013)		-.026 (±.013)	-.055* (±.014)	-.078** (±.013)	-.100** (±.015)	-.109** (±.015)	-.115** (±.016)
6	.136** (±.026)	.123** (±.024)	.131** (±.022)	.078** (±.015)	.026 (±.013)		-.029 (±.010)	-.052** (±.011)	-.074** (±.014)	-.084** (±.015)	-.089** (±.015)
7	.165** (±.026)	.152** (±.025)	.160** (±.022)	.107** (±.016)	.055* (±.014)	.029 (±.010)		-.023 (±.008)	-.045** (±.010)	-.054** (±.010)	-.060** (±.011)
8	.188** (±.025)	.175** (±.023)	.183** (±.023)	.130** (±.015)	.078** (±.013)	.052** (±.011)	.023 (±.008)		-.022 (±.007)	-.032* (±.008)	-.037** (±.009)
9	.210** (±.025)	.197** (±.024)	.205** (±.024)	.152** (±.017)	.100** (±.015)	.074** (±.014)	.045** (±.010)	.022 (±.007)		-.010 (±.004)	-.015* (±.004)
10	.219** (±.025)	.207** (±.024)	.215** (±.024)	.162** (±.017)	.109** (±.015)	.084** (±.015)	.054** (±.010)	.032* (±.008)	.010 (±.004)		-.005 (±.004)
11	.225** (±.025)	.212** (±.023)	.22** (±.023)	.167** (±.017)	.115** (±.016)	.089** (±.015)	.060** (±.011)	.037** (±.009)	.015* (±.004)	.005 (±.004)	

**P<.01 * P<.05

Pairwise comparison table depicting results based on the results of a one-way ANOVA depicting the relationship between Frequency Error and the nSweeps condition for the hidden component of the adult data at ten iterations.

FE	1	2	3	4	5	6	7	8	9	10	11
1		.343 (±.449)	.727 (±.647)	2.183 (±.669)	3.542** (±.653)	3.909** (±.695)	4.239** (±.670)	4.484** (±.635)	4.653** (±.622)	4.708** (±.636)	4.796** (±.623)
2	-.343 (±.449)		.384 (±.455)	1.840 (±.507)	3.199** (±.556)	3.566** (±.594)	3.896** (±.609)	4.141** (±.574)	4.310** (±.555)	4.365** (±.583)	4.453** (±.562)
3	-.727 (±.647)	-.384 (±.455)		1.456* (±.350)	2.815** (±.481)	3.182** (±.527)	3.512** (±.525)	3.757** (±.547)	3.926** (±.541)	3.981** (±.551)	4.069** (±.551)
4	-2.183 (±.669)	-1.840 (±.507)	-1.456* (±.350)		1.359** (±.262)	1.726** (±.301)	2.057** (±.333)	2.301** (±.367)	2.470** (±.355)	2.525** (±.363)	2.614** (±.370)
5	-3.542** (±.653)	-3.199** (±.556)	-2.815** (±.481)	-1.359** (±.262)		.367 (±.159)	.697 (±.195)	.942** (±.205)	1.111** (±.237)	1.166** (±.229)	**1.255 (±.250)
6	-3.909** (±.695)	-3.566** (±.594)	-3.182** (±.527)	-1.726** (±.301)	-.367 (±.159)		.330 (±.108)	.575 (±.130)	.744** (±.152)	.799** (±.148)	.88**8 (±.172)
7	-4.239** (±.670)	-3.896** (±.609)	-3.512** (±.525)	-2.057** (±.333)	-.697 (±.195)	-.330 (±.108)		.244 (±.099)	.414 (±.115)	.469** (±.091)	.557** (±.126)
8	-4.484** (±.635)	-4.141** (±.574)	-3.757** (±.547)	-2.301** (±.367)	-.942** (±.205)	-.575** (±.130)	-.244 (±.099)		.169 (±.104)	.224 (±.088)	.313 (±.098)
9	-4.653** (±.622)	-4.310** (±.555)	-3.926** (±.541)	-2.470** (±.355)	-1.111** (±.237)	-.744** (±.152)	-.414 (±.115)	-.169 (±.104)		.055 (±.053)	.144 (±.059)
10	-4.708** (±.636)	-4.365** (±.583)	-3.981** (±.551)	-2.525** (±.363)	-1.166** (±.229)	-.799** (±.148)	-.469** (±.091)	-.224 (±.088)	-.055 (±.053)		.088 (±.062)
11	-4.796** (±.623)	-4.453** (±.562)	-4.069** (±.551)	-2.614** (±.370)	-1.255** (±.250)	-.888** (±.172)	-.557** (±.126)	-.313 (±.098)	-.144 (±.059)	-.088 (±.062)	

**P<.01 * P<.05

Pairwise comparison table depicting results based on the results of a one-way ANOVA depicting the relationship between Slope Error and the nSweeps condition for the hidden component of the adult data at ten iterations.

SE	1	2	3	4	5	6	7	8	9	10	11
1		-25.422 (±7.256)	- 45.997** (±8.765)	- 70.801** (±7.214)	- 84.244** (±9.460)	- 90.045** (±10.327)	- 96.588** (±8.987)	- 97.647** (±9.718)	- 100.128** (±9.513)	- 100.976** (±9.581)	- 101.682** (±9.445)
2	25.422 (±7.256)		-20.575 (±9.337)	- 45.379** (±7.673)	- 58.822** (±10.506)	- 64.623** (±11.253)	- 71.166** (±10.997)	- 72.225** (±10.514)	-74.706** (±11.210)	-75.554** (±11.313)	-76.260** (±11.383)
3	45.997** (±8.765)	20.575 (±9.337)		-24.804 (±6.622)	- 38.247** (±8.517)	- 44.048** (±8.315)	- 50.592** (±8.194)	- 51.650** (±8.527)	-54.131** (±9.420)	-54.979** (±9.708)	-55.685** (±9.826)
4	70.801** (±7.214)	45.379** (±7.673)	24.804 (±6.622)		-13.443 (±4.218)	-19.244* (±5.026)	- 25.788** (±4.883)	- 26.846** (±5.272)	-29.327** (±5.741)	-30.175** (±5.827)	-30.881** (±6.025)
5	84.244** (±9.460)	58.822** (±10.506)	38.247** (±8.517)	13.443 (±4.218)		-5.801 (±3.755)	-12.344 (±3.966)	-13.403 (±4.894)	-15.884 (±5.171)	-16.732 (±5.041)	-17.438 (±5.737)
6	90.045** (±10.327)	64.623** (±11.253)	44.048** (±8.315)	19.244* (±5.026)	5.801 (±3.755)		-6.543 (±3.112)	-7.602 (±2.739)	-10.083 (±4.319)	-10.931 (±4.631)	-11.637 (±5.116)
7	96.588** (±8.987)	71.166** (±10.997)	50.592** (±8.194)	25.788** (±4.883)	12.344 (±3.966)	6.543 (±3.112)		-1.058 (±2.693)	-3.539 (±2.441)	-4.388 (±2.789)	-5.094 (±3.172)
8	97.647** (±9.718)	72.225** (±10.514)	51.650** (±8.527)	26.846** (±5.272)	13.403 (±4.894)	7.602 (±2.739)	1.058 (±2.693)		-2.481 (±3.199)	-3.329 (±3.714)	-4.035 (±4.001)
9	100.128** (±9.513)	74.706** (±11.210)	54.131** (±9.420)	29.327** (±5.741)	15.884 (±5.171)	10.083 (±4.319)	3.539 (±2.441)	2.481 (±3.199)		-.848 (±1.113)	-1.554 (±1.284)
10	100.976** (±9.581)	75.554** (±11.313)	54.979** (±9.708)	30.175** (±5.827)	16.732 (±5.041)	10.931 (±4.631)	4.388 (±2.789)	3.329 (±3.714)	.848 (±1.113)		-.706 (±1.086)
11	101.682** (±9.445)	76.260** (±11.383)	55.685** (±9.826)	30.881** (±6.025)	17.438 (±5.737)	11.637 (±5.116)	5.094 (±3.172)	4.035 (±4.001)	1.554 (±1.284)	.706 (±1.086)	

**P<.01 * P<.05

Pairwise comparison table depicting results based on the results of a one-way ANOVA depicting the relationship between Tracking Accuracy and the nSweeps condition for the hidden component of the adult data at ten iterations.

TA	1	2	3	4	5	6	7	8	9	10	11
1		-.007 (±.029)	-.022 (±.042)	-.111 (±.041)	-.178* (±.041)	-.192* (±.045)	-.217** (±.041)	-.222** (±.042)	-.234** (±.042)	-.239** (±.041)	-.240** (±.041)
2	.007 (±.029)		-.015 (±.028)	-.104* (±.026)	-.172** (±.031)	-.185** (±.033)	-.210** (±.031)	-.215** (±.031)	-.227** (±.034)	-.232** (±.035)	-.234** (±.034)
3	.022 (±.042)	.015 (±.028)		-.089* (±.022)	-.157** (±.033)	-.170** (±.035)	-.195** (±.034)	-.200** (±.037)	-.212** (±.038)	-.217** (±.038)	-.219** (±.039)
4	.111 (±.041)	.104* (±.026)	.089* (±.022)		-.068 (±.019)	-.081* (±.021)	-.106** (±.021)	-.111** (±.024)	-.123** (±.025)	-.128** (±.026)	-.130** (±.026)
5	.178* (±.041)	.172** (±.031)	.157** (±.033)	.068 (±.019)		-.014 (±.014)	-.038 (±.016)	-.043 (±.019)	-.055 (±.021)	-.060 (±.021)	-.062 (±.023)
6	.192* (±.045)	.185** (±.033)	.170** (±.035)	.081* (±.021)	.014 (±.014)		-.025 (±.010)	-.030 (±.011)	-.042 (±.014)	-.047 (±.015)	-.048 (±.015)
7	.217** (±.041)	.210** (±.031)	.195** (±.034)	.106** (±.021)	.038 (±.016)	.025 (±.010)		-.005 (±.009)	-.017 (±.010)	-.022 (±.011)	-.024 (±.011)
8	.222** (±.042)	.215** (±.031)	.200** (±.037)	.111** (±.024)	.043 (±.019)	.030 (±.011)	.005 (±.009)		-.012 (±.008)	-.017 (±.009)	-.019 (±.008)
9	.234** (±.042)	.227** (±.034)	.212** (±.038)	.123** (±.025)	.055 (±.021)	.042 (±.014)	.017 (±.010)	.012 (±.008)		-.005 (±.004)	-.007 (±.004)
10	.239** (±.041)	.232** (±.035)	.217** (±.038)	.128** (±.026)	.060 (±.021)	.047 (±.015)	.022 (±.011)	.017 (±.009)	.005 (±.004)		-.002 (±.004)
11	.240** (±.041)	.234** (±.034)	.219** (±.039)	.30** (±.026)	.062 (±.023)	.048 (±.015)	.024 (±.011)	.019 (±.008)	.007 (±.004)	.002 (±.004)	

**P<.01 * P<.05

Pairwise comparison table depicting results based on the results of a one-way ANOVA depicting the relationship between Frequency Error and the nSweeps condition for the hidden component of the newborn data at ten iterations.

FE	1	2	3	4	5	6	7	8	9	10	11
1		.381 (±.357)	.464 (±.292)	1.210 (±.350)	2.310** (±.356)	2.973** (±.372)	3.644** (±.459)	4.194** (±.418)	4.785** (±.372)	4.862** (±.400)	5.050** (±.396)
2	-.381 (±.357)		.083 (±.387)	.829 (±.422)	1.929** (±.394)	2.592** (±.441)	3.263** (±.516)	3.813** (±.492)	4.404** (±.460)	4.481** (±.468)	4.669** (±.465)
3	-.464 (±.292)	-.083 (±.387)		.746 (±.310)	1.847** (±.366)	2.509** (±.379)	3.180** (±.429)	3.730** (±.356)	4.322** (±.348)	4.398** (±.370)	4.586** (±.369)
4	-1.210 (±.350)	-.829 (±.422)	-.746 (±.310)		1.101** (±.221)	1.763** (±.291)	2.434** (±.312)	2.984** (±.223)	3.576** (±.218)	3.652** (±.241)	3.840** (±.269)
5	-2.310** (±.356)	-1.929** (±.394)	-1.847** (±.366)	-1.101** (±.221)		.663 (±.218)	1.333** (±.272)	1.884** (±.232)	2.475** (±.231)	2.552** (±.238)	2.740** (±.262)
6	-2.973** (±.372)	-2.592** (±.441)	-2.509** (±.379)	-1.763** (±.291)	-.663 (±.218)		.671 (±.222)	1.221** (±.223)	1.813** (±.236)	1.889** (±.263)	2.077** (±.277)
7	-3.644** (±.459)	-3.263** (±.516)	-3.180** (±.429)	-2.434** (±.312)	-1.333** (±.272)	-.671 (±.222)		.551 (±.186)	1.142** (±.193)	1.218** (±.179)	1.406** (±.203)
8	-4.194** (±.418)	-3.813** (±.492)	-3.730** (±.356)	-2.984** (±.223)	-1.884** (±.232)	-1.221** (±.223)	-.551 (±.186)		.591 (±.146)	.668** (±.161)	.856** (±.194)
9	-4.785** (±.372)	-4.404** (±.460)	-4.322** (±.348)	-3.576** (±.218)	-2.475** (±.231)	-1.813** (±.236)	-1.142** (±.193)	-.591* (±.146)		.077* (±.074)	.265 (±.094)
10	-4.862** (±.400)	-4.481** (±.468)	-4.398** (±.370)	-3.652** (±.241)	-2.552** (±.238)	-1.889** (±.263)	-1.218** (±.179)	-.668* (±.161)	-.077 (±.074)		.188 (±.070)
11	-5.050** (±.396)	-4.669** (±.465)	-4.586** (±.369)	-3.840** (±.269)	-2.740** (±.262)	-2.077** (±.277)	-1.406** (±.203)	-.856** (±.194)	-.265 (±.094)	-.188 (±.070)	

**P<.01 * P<.05

247

Pairwise comparison table depicting results based on the results of a one-way ANOVA depicting the relationship between Slope Error and the nSweeps condition for the hidden component of the newborn data at ten iterations.

SE	1	2	3	4	5	6	7	8	9	10	11
1		-4.770 (±7.312)	-18.501 (±8.820)	- 41.635** (±9.194)	-63.678** (±10.095)	-73.946** (±11.734)	-87.099** (±12.040)	-91.574** (±12.316)	- 100.696** (±12.702)	- 103.315** (±12.593)	- 105.202** (±12.266)
2	4.770 (±7.312)		-13.731 (±8.513)	-36.865* (±8.843)	-58.908** (±7.693)	-69.176** (±8.383)	-82.330** (±8.846)	-86.804** (±8.840)	-95.926** (±9.537)	-98.545** (±9.586)	- 100.432** (±9.222)
3	18.501 (±8.820)	13.731 (±8.513)		-23.133 (±8.407)	-45.176** (±6.780)	-55.445** (±7.606)	-68.598** (±7.772)	-73.073** (±8.289)	-82.195** (±8.656)	-84.813** (±8.303)	-86.701** (±8.354)
4	41.635** (±9.194)	36.865* (±8.843)	23.133 (±8.407)		-22.043 (±6.648)	-32.311* (±8.496)	-45.465** (±8.037)	-49.939** (±8.291)	-59.062** (±9.029)	-61.680** (±8.759)	-63.567** (±8.944)
5	63.678** (±10.095)	58.908** (±7.693)	45.176** (±6.780)	22.043 (±6.648)		-10.268 (±5.425)	-23.422* (±5.486)	-27.896** (±6.279)	-37.019** (±7.346)	-39.637** (±7.417)	-41.524** (±7.608)
6	73.946** (±11.734)	69.176** (±8.383)	55.445** (±7.606)	32.311* (±8.496)	10.268 (±5.425)		-13.154 (±3.943)	-17.628 (±5.097)	-26.750** (±6.151)	-29.369** (±6.338)	-31.256** (±6.374)
7	87.099** (±12.040)	82.330** (±8.846)	68.598** (±7.772)	45.465** (±8.037)	23.422* (±5.486)	13.154 (±3.943)		-4.474 (±3.466)	-13.597 (±3.963)	-16.215* (±3.976)	-18.102** (±4.034)
8	91.574** (±12.316)	86.804** (±8.840)	73.073** (±8.289)	49.939** (±8.291)	27.896** (±6.279)	17.628 (±5.097)	4.474 (±3.466)		-9.122 (±3.110)	-11.741 (±3.467)	-13.628 (±3.866)
9	100.696** (±12.702)	95.926** (±9.537)	82.195** (±8.656)	59.062** (±9.029)	37.019** (±7.346)	26.75** (±6.151)	13.597 (±3.963)	9.122 (±3.110)		-2.618 (±2.117)	-4.506 (±1.905)
10	103.315** (±12.593)	98.545** (±9.586)	84.813** (±8.303)	61.680** (±8.759)	39.637** (±7.417)	29.369** (±6.338)	16.215* (±3.976)	11.741 (±3.467)	2.618 (±2.117)		-1.887 (±1.766)
11	105.202** (±12.266)	100.432** (±9.222)	86.701** (±8.354)	63.567** (±8.944)	41.524** (±7.608)	31.256** (±6.374)	18.102** (±4.034)	13.628 (±3.866)	4.506 (±1.905)	1.887 (±1.766)	

**P<.01 * P<.05

Pairwise comparison table depicting results based on the results of a one-way ANOVA depicting the relationship between Tracking Accuracy and the nSweeps condition for the hidden component of the newborn data at ten iterations.

TA	1	2	3	4	5	6	7	8	9	10	11
1		-.019 (±.030)	.012 (±.027)	-.005 (±.019)	-.042 (±.025)	-.080 (±.026)	-.113 (±.032)	-.154** (±.029)	-.186** (±.030)	-.200** (±.030)	-.209** (±.030)
2	.019 (±.030)		.031 (±.027)	.014 (±.031)	-.023 (±.033)	-.061 (±.035)	-.095 (±.038)	-.135* (±.035)	-.167** (±.034)	-.181** (±.034)	-.190** (±.032)
3	-.012 (±.027)	-.031 (±.027)		-.017 (±.023)	-.054 (±.027)	-.092 (±.026)	-.126** (±.029)	-.166** (±.027)	-.198** (±.029)	-.212** (±.028)	-.221** (±.027)
4	.005 (±.019)	-.014 (±.031)	.017 (±.023)		-.038 (±.018)	-.075 (±.022)	-.109** (±.024)	-.149** (±.020)	-.182** (±.022)	-.196** (±.022)	-.205** (±.021)
5	.042 (±.025)	.023 (±.033)	.054 (±.027)	.038 (±.018)		-.038 (±.021)	-.071 (±.023)	-.112** (±.019)	-.144** (±.021)	-.158** (±.022)	-.167** (±.022)
6	.080 (±.026)	.061 (±.035)	.092 (±.026)	.075 (±.022)	.038 (±.021)		-.034 (±.018)	-.074* (±.019)	-.106** (±.024)	-.120** (±.025)	-.129** (±.024)
7	.113 (±.032)	.095 (±.038)	.126** (±.029)	.109 (±.024)	.071 (±.023)	.034 (±.018)		-.040 (±.013)	-.073* (±.017)	-.087** (±.017)	-.096** (±.017)
8	.154** (±.029)	.135* (±.035)	.166** (±.027)	.149** (±.020)	.112** (±.019)	.074* (±.019)	.040 (±.013)		-.032 (±.013)	-.046 (±.013)	-.055* (±.015)
9	.186** (±.030)	.167** (±.034)	.198** (±.029)	.182** (±.022)	.144** (±.021)	.106** (±.024)	.073* (±.017)	.032 (±.013)		-.014 (±.007)	-.023 (±.006)
10	.200** (±.030)	.181** (±.034)	.212** (±.028)	.196** (±.022)	.158** (±.022)	.120** (±.025)	.087** (±.017)	.046 (±.013)	.014 (±.007)		-.009 (±.006)
11	.209** (±.030)	.190** (±.032)	.221** (±.027)	.205** (±.021)	.167** (±.022)	.129** (±.024)	.096** (±.017)	.055* (±.015)	.023 (±.006)	.009 (±.006)	

**P<.01 * P<.05

Pairwise comparison table depicting results based on the results of a two-way ANOVA depicting the relationship between Frequency Error and the nSweeps condition for the newborn and adult data at ten iterations.

FE	1	2	3	4	5	6	7	8	9	10	11
1		-.967* (±.251)	-2.340** (±.254)	-3.544** (±.298)	-5.017** (±.337)	-5.873** (±.302)	-6.304** (±.314)	-6.649** (±.325)	-6.906** (±.327)	-7.156** (±.309)	-7.330** (±.319)
2	.967* (±.251)		-1.372** (±.203)	-2.577** (±.214)	-4.049** (±.267)	-4.905** (±.272)	-5.337** (±.300)	-5.682** (±.321)	-5.939** (±.326)	-6.188** (±.325)	-6.362** (±.337)
3	2.340** (±.254)	1.372** (±.203)		-1.204** (±.167)	-2.677** (±.221)	-3.533** (±.210)	-3.965** (±.241)	-4.309** (±.252)	-4.566** (±.274)	-4.816** (±.273)	-4.990** (±.286)
4	3.544** (±.298)	2.577** (±.214)	1.204** (±.167)		-1.473** (±.151)	-2.329** (±.166)	-2.760** (±.207)	-3.105** (±.231)	-3.362** (±.252)	-3.612** (±.254)	-3.786** (±.277)
5	5.017** (±.337)	4.049** (±.267)	2.677** (±.221)	1.473** (±.151)		-.856** (±.116)	-1.288** (±.158)	-1.632** (±.182)	-1.889** (±.202)	-2.139** (±.213)	-2.313** (±.242)
6	5.873** (±.302)	4.905** (±.272)	3.533** (±.210)	2.329** (±.166)	.856** (±.116)		-.432* (±.113)	-.776** (±.137)	-1.033** (±.163)	-1.283** (±.164)	-1.457** (±.190)
7	6.304** (±.314)	5.337** (±.300)	3.965** (±.241)	2.760** (±.207)	1.288** (±.158)	.432* (±.113)		-.345 (±.107)	-.601** (±.096)	-.851** (±.109)	-1.025** (±.131)
8	6.649** (±.325)	5.682** (±.321)	4.309** (±.252)	3.105** (±.231)	1.632** (±.182)	.776** (±.137)	.345 (±.107)		-.257 (±.086)	-.507 (±.101)	-.681** (±.116)
9	6.906** (±.327)	5.939** (±.326)	4.566** (±.274)	3.362** (±.252)	1.889** (±.202)	1.033** (±.163)	.601** (±.096)	.257 (±.086)		-.250 (±.072)	-.424** (±.080)
10	7.156** (±.309)	6.188** (±.325)	4.816** (±.273)	3.612** (±.254)	2.139** (±.213)	1.283** (±).164	.851** (±.109)	.507** (±.101)	.250 (±.072)		-.174 (±.057)
11	7.330** (±.319)	6.362** (±.337)	4.990** (±.286)	3.786** (±.277)	2.313** (±.242)	1.457** (±.190)	1.025** (±.131)	.681** (±.116)	.424** (±.080)	.174 (±.057)	

**P<.01 * P<.05

Pairwise comparison table depicting results based on the results of a two-way ANOVA depicting the relationship between Slope Error and the nSweeps condition for the newborn and adult data at ten iterations.

SE	1	2	3	4	5	6	7	8	9	10	11
1		17.864 (±5.122)	46.067** (±6.229)	60.939** (±5.996)	80.747** (±5.748)	93.679** (±6.034)	97.780** (±6.428)	103.433** (±6.161)	108.272** (±6.668)	110.997** (±6.691)	113.262** (±6.957)
2	-17.864 (±5.122)		28.203** (±6.370)	43.075** (±5.773)	62.883** (±5.314)	75.815** (±6.527)	79.916** (±6.931)	85.569** (±6.833)	90.408** (±6.853)	93.133** (±6.945)	95.398** (±7.178)
3	-46.067** (±6.229)	-28.203** (±6.370)		14.872 (±6.248)	34.681** (±5.176)	47.612** (±5.456)	51.713** (±6.030)	57.366** (±6.461)	62.205** (±6.485)	64.930** (±6.711)	67.196** (±7.097)
4	-60.939** (±5.996)	-43.075** (±5.773)	-14.872 (±6.248)		19.809** (±4.603)	32.740** (±5.277)	36.842** (±5.518)	42.495** (±5.241)	47.333** (±5.797)	50.058** (±5.941)	52.324** (±6.082)
5	-80.747** (±5.748)	-62.883** (±5.314)	-34.681** (±5.176)	-19.809** (±4.603)		12.931 (±3.775)	17.033* (±4.465)	22.686** (±4.712)	27.525** (±4.862)	30.250** (±5.024)	32.515** (±5.448)
6	-93.679** (±6.034)	-75.815** (±6.527)	-47.612** (±5.456)	-32.740** (±5.277)	-12.931 (±3.775)		4.102 (±2.973)	9.755 (±3.369)	14.594** (±3.249)	17.319** (±3.363)	19.584** (±3.886)
7	-97.780** (±6.428)	-79.916** (±6.931)	-51.713** (±6.030)	-36.842** (±5.518)	-17.033* (±4.465)	-4.102 (±2.973)		5.653 (±2.895)	10.492* (±2.634)	13.217** (±3.133)	15.482** (±3.156)
8	-103.433** (±6.161)	-85.569** (±6.833)	-57.366** (±6.461)	-42.495** (±5.241)	-22.686** (±4.712)	-9.755 (±3.369)	-5.653 (±2.895)		4.839 (±2.505)	7.564 (±2.401)	9.829* (±2.689)
9	-108.272** (±6.668)	-90.403** (±6.853)	-62.205** (±6.485)	-47.333** (±5.797)	-27.525** (±4.862)	-14.594** (±3.249)	-10.492* (±2.634)	-4.839 (±2.505)		2.725 (±1.583)	4.990 (±1.708)
10	-110.997** (±6.691)	-93.133** (±6.945)	-64.930** (±6.711)	-50.058** (±5.941)	-30.250** (±5.024)	-17.319** (±3.363)	-13.217** (±3.133)	-7.564 (±2.401)	-2.725 (±1.583)		2.265 (±1.158)
11	-113.262** (±6.957)	-95.398** (±7.178)	-67.196** (±7.097)	-52.324** (±6.082)	-32.515** (±5.448)	-19.584** (±3.886)	-15.482** (±3.156)	-9.829* (±2.689)	-4.990 (±1.708)	-2.265 (±1.158)	

**P<.01 * P<.05

Pairwise comparison table depicting results based on the results of a two-way ANOVA depicting the relationship between Tracking Accuracy and the nSweeps condition for the newborn and adult data at ten iterations.

TA	1	2	3	4	5	6	7	8	9	10	11
1		.052* (±.013)	.120** (±.015)	.180** (±.017)	.255** (±.021)	.306** (±.018)	.333** (±.018)	.355** (±.018)	.374** (±.018)	.389** (±.017)	.399** (±.018)
2	-.052* (±.013)		.067** (±.013)	.128** (±.015)	.203** (±.017)	.254** (±.019)	.281** (±.019)	.303** (±.020)	.322** (±.020)	.337** (±.019)	.347** (±.021)
3	-.120** (±.015)	-.067** (±.013)		.061** (±.013)	.135** (±.014)	.186** (±.015)	.214** (±.015)	.236** (±.016)	.254** (±.017)	.269** (±.017)	.280** (±.018)
4	-.180** (±.017)	-.128** (±.015)	-.061** (±.013)		.075** (±.012)	.126** (±.014)	.153** (±.014)	.175** (±.015)	.193** (±.016)	.208** (±.016)	.219** (±.017)
5	-.255** (±.021)	-.203** (±.017)	-.135** (±.014)	-.075** (±.012)		.051** (±.012)	.078** (±.014)	.100** (±.016)	.119** (±.016)	.134** (±.017)	.144** (±.019)
6	-.306** (±.018)	-.254** (±.019)	-.186** (±.015)	-.126** (±.014)	-.051** (±.012)		.027 (±.009)	.049** (±.010)	.068** (±.012)	.083** (±.011)	.093** (±.014)
7	-.333** (±.018)	-.281** (±.019)	-.214** (±.015)	-.153** (±.014)	-.078** (±.014)	-.027 (±.009)		.022 (±.010)	.041** (±.008)	.055** (±.009)	.066** (±.010)
8	-.355** (±.018)	-.303** (±.020)	-.236** (±.016)	-.175** (±.015)	-.100** (±.016)	-.049** (±.010)	-.022 (±.010)		.019 (±.009)	.034 (±.007)	.044** (±.009)
9	-.374** (±.018)	-.322** (±.020)	-.254** (±.017)	-.193** (±.016)	-.119** (±.016)	-.068** (±.012)	-.041** (±.008)	-.019 (±.009)		.015 (±.006)	.025** (±.006)
10	-.389** (±.017)	-.337** (±.019)	-.269** (±.017)	-.208** (±.016)	-.134** (±.017)	-.083** (±.011)	-.055** (±.009)	-.034** (±.007)	-.015 (±.006)		.010 (±.005)
11	-.399** (±.018)	-.347** (±.021)	-.280** (±.018)	-.219** (±.017)	-.144** (±.019)	-.093** (±.014)	-.066** (±.010)	-.044** (±.009)	-.025** (±.006)	-.010 (±.005)	

**P<.01 * P<.05

Pairwise comparison table depicting results based on the results of a two-way ANOVA depicting the relationship between Frequency Error and the nSweeps condition for the adult data at ten iterations.

FE	1	2	3	4	5	6	7	8	9	10	11
1		-1.487* (±.382)	-3.198** (±.407)	-4.785** (±.415)	-5.903** (±.448)	-6.566** (±.429)	-6.789** (±.443)	-7.077** (±.458)	-7.200** (±.455)	-7.482** (±.432)	-7.513** (±.435)
2	1.487* (±.382)		-1.711** (±.300)	-3.298** (±.329)	-4.416** (±.402)	-5.079** (±.457)	-5.302** (±.483)	-5.590** (±.512)	-5.714** (±.515)	-5.995** (±.517)	-6.026** (±.527)
3	3.198** (±.407)	1.711** (±.300)		-1.587** (±.195)	-2.704** (±.290)	-3.367** (±.317)	-3.590** (±.359)	-3.879** (±.371)	-4.002** (±.396)	-4.284** (±.403)	-4.314** (±.412)
4	4.785** (±.415)	3.298** (±.329)	1.587** (±.195)		-1.117** (±.183)	-1.781** (±.241)	-2.004** (±.275)	-2.292** (±.307)	-2.415** (±.324)	-2.697** (±.343)	-2.728** (±.354)
5	5.903** (±.448)	4.416** (±.402)	2.704** (±.290)	1.117** (±.183)		-.663* (±.156)	-.886** (±.197)	-1.175** (±.230)	-1.298** (±.226)	-1.580** (±.246)	-1.610** (±.261)
6	6.566** (±.429)	5.079** (±.457)	3.367** (±.317)	1.781** (±.241)	.663* (±.156)		-.223 (±.100)	-.511** (±.115)	-.635** (±.147)	-.916** (±.163)	-.947** (±.174)
7	6.789** (±.443)	5.302** (±.483)	3.590** (±.359)	2.004** (±.275)	.886** (±.197)	.223 (±.100)		-.288 (±.103)	-.412* (±.099)	-.694** (±.128)	-.724** (±.138)
8	7.077** (±.458)	5.590** (±.512)	3.879** (±.371)	2.292** (±.307)	1.175** (±.230)	.511** (±.115)	.288 (±.103)		-.123 (±.107)	-.405 (±.120)	-.436 (±).120
9	7.200** (±.455)	5.714** (±.515)	4.002** (±.396)	2.415** (±.324)	1.298** (±.226)	.635** (±.147)	.412* (±.099)	.123 (±.107)		-.282* (±.074)	-.312 (±.085)
10	7.482** (±.432)	5.995** (±.517)	4.284** (±.403)	2.697** (±.343)	1.580** (±.246)	.916** (±.163)	.694** (±.128)	.405 (±.120)	.282* (±.074)		-.030 (±.032)
11	7.513** (±.435)	6.026** (±.527)	4.314** (±.412)	2.728** (±.354)	1.610** (±.261)	.947** (±.174)	.724** (±.138)	.436 (±.120)	.312 (±.085)	.030 (±.032)	

**P<.01 * P<.05

253

Pairwise comparison table depicting results based on the results of a two-way ANOVA depicting the relationship between Slope Error and the nSweeps condition for the adult data at ten iterations.

SE	1	2	3	4	5	6	7	8	9	10	11
1		29.274** (±5.580)	64.703** (±6.590)	83.176** (±6.680)	88.957** (±7.230)	100.237** (±8.618)	101.641** (±8.778)	107.852** (±9.514)	107.786** (±10.189)	110.288** (±10.850)	111.826** (±10.927)
2	-29.274** (±5.580)		35.429** (±6.570)	53.901** (±7.031)	59.683** (±8.048)	70.963** (±9.302)	72.367** (±9.670)	78.578** (±10.428)	78.512** (±10.876)	81.014** (±11.359)	82.552** (±11.477)
3	-64.703** (±6.590)	-35.429** (±6.570)		18.473 (±5.993)	24.255* (±5.938)	35.534** (±6.672)	36.939** (±7.356)	43.150** (±8.307)	43.084** (±8.738)	45.586** (±9.468)	47.123** (±9.701)
4	-83.176** (±6.680)	-53.901** (±7.031)	-18.473 (±5.993)		5.782 (±4.632)	17.061 (±6.131)	18.466 (±5.840)	24.677 (±6.997)	24.611 (±7.148)	27.113 (±8.245)	28.650 (±8.271)
5	-88.957** (±7.230)	-59.683** (±8.048)	-24.255* (±5.938)	-5.782 (±4.632)		11.280 (±4.179)	12.684 (±4.083)	18.895* (±5.007)	18.829 (±5.394)	21.331 (±6.399)	22.868 (±6.633)
6	- 100.237** (±8.618)	-70.963** (±9.302)	- 35.534** (±6.672)	-17.061 (±6.131)	-11.280 (±4.179)		1.404 (±2.594)	7.615 (±2.753)	7.549 (±3.518)	10.051 (±3.902)	11.589 (±4.086)
7	- 101.641** (±8.778)	-72.367** (±9.670)	- 36.939** (±7.356)	-18.466 (±5.840)	-12.684 (±4.083)	-1.404 (±2.594)		6.211 (±2.993)	6.145 (±3.340)	8.647 (±3.844)	10.185 (±4.142)
8	- 107.852** (±9.514)	-78.578** (±10.428)	- 43.150** (±8.307)	-24.677 (±6.997)	-18.895* (±5.007)	-7.615 (±2.753)	-6.211 (±2.993)		-.066 (±2.590)	2.436 (±2.748)	3.973 (±2.871)
9	- 107.786** (±10.189)	-78.512** (±10.876)	- 43.084** (±8.738)	-24.611 (±7.148)	-18.829 (±5.394)	-7.549 (±3.518)	-6.145 (±3.340)	.066 (±2.590)		2.502 (±1.950)	4.040 (±1.809)
10	- 110.288** (±10.850)	-81.014** (±11.359)	- 45.586** (±9.468)	-27.113 (±8.245)	-21.331 (±6.399)	-10.051 (±3.902)	-8.647 (±3.844)	-2.436 (±2.748)	-2.502 (±1.950)		1.538 (±.862)
11	- 111.826** (±10.927)	-82.552** (±11.477)	- 47.123** (±9.701)	-28.650 (±8.271)	-22.868 (±6.633)	-11.589 (±4.086)	-10.18 (±4.142)	-3.973 (±2.871)	-4.040 (±1.809)	-1.538 (±.862)	

**P<.01 * P<.05

Pairwise comparison table depicting results based on the results of a two-way ANOVA depicting the relationship between Tracking Accuracy and the nSweeps condition for the adult data at ten iterations.

TA	1	2	3	4	5	6	7	8	9	10	11
1		.084* (±.020)	.171** (±.022)	.254** (±.023)	.289** (±.028)	.337** (±.023)	.348** (±.024)	.361** (±.026)	.368** (±.026)	.385** (±.024)	.387** (±.024)
2	-.084* (±.020)		.087** (±.017)	.171** (±.020)	.205** (±.025)	.253** (±.026)	.264** (±.028)	.277** (±.030)	.284** (±.030)	.301** (±.030)	.304** (±.030)
3	-.171** (±.022)	-.087** (±.017)		.084** (±.017)	.118** (±.019)	.166** (±.020)	.177** (±.023)	.190** (±.025)	.197** (±.025)	.214** (±.025)	.216** (±.026)
4	-.254** (±.023)	-.171** (±.020)	-.084** (±.017)		.034 (±.014)	.083** (±.018)	.093** (±.019)	.106** (±.023)	.113** (±.022)	.131** (±.023)	.133** (±.024)
5	-.289** (±.028)	-.205** (±.025)	-.118** (±.019)	-.034 (±.014)		.049 (±.015)	.059 (±.017)	.072* (±.019)	.079** (±.018)	.097** (±.020)	.099** (±.022)
6	-.337** (±.023)	-.253** (±.026)	-.166** (±.020)	-.083** (±.018)	-.049 (±.015)		.010 (±.009)	.024 (±.008)	.030 (±.011)	.048** (±.011)	.050* (±.012)
7	-.348** (±.024)	-.264** (±.028)	-.177** (±.023)	-.093** (±.019)	-.059 (±.017)	-.010 (±.009)		.013 (±.009)	.020 (±.008)	.038* (±.010)	.040* (±.010)
8	-.361** (±.026)	-.277** (±.030)	-.190** (±.025)	-.106** (±.023)	-.072* (±.019)	-.024 (±.008)	-.013 (±.009)		.007 (±.009)	.024 (±.010)	.027 (±.010)
9	-.368** (±.026)	-.284** (±.030)	-.197** (±.025)	-.113** (±.022)	-.079** (±.018)	-.030 (±.011)	-.020 (±.008)	-.007 (±.009)		.018 (±.006)	.020 (±.007)
10	-.385** (±.024)	-.301** (±.030)	-.214** (±.025)	-.131** (±.023)	-.097** (±.020)	-.048** (±.011)	-.038* (±.010)	-.024 (±.010)	-.018 (±.006)		.002 (±.002)
11	-.387** (±.024)	-.304** (±.030)	-.216** (±.026)	-.133** (±.024)	-.099** (±.022)	-.050* (±.012)	-.040* (±.010)	-.027 (±.010)	-.020 (±.007)	-.002 (±.002)	

**P<.01 * P<.05

Pairwise comparison table depicting results based on the results of a two-way ANOVA depicting the relationship between Frequency Error and the nSweeps condition for the newborn data at ten iterations.

FE	1	2	3	4	5	6	7	8	9	10	11
1		-.448 (±.325)	-1.481** (±.304)	-2.303** (±.429)	-4.131** (±.503)	-5.179** (±.424)	-5.820** (±.444)	-6.221** (±.462)	-6.611** (±.471)	-6.829** (±.443)	-7.147** (±.468)
2	.448 (±.325)		-1.033* (±.272)	-1.855** (±.273)	-3.683** (±.353)	-4.732** (±.295)	-5.372** (±.357)	-5.773** (±.388)	-6.163** (±.401)	-6.381** (±.394)	-6.699** (±.420)
3	1.481** (±.304)	1.033* (±.272)		-.822 (±.272)	-2.650** (±.335)	-3.698** (±.276)	-4.339** (±.321)	-4.740** (±.341)	-5.130** (±.380)	-5.348** (±.369)	-5.666** (±.395)
4	2.303** (±.429)	1.855** (±.273)	.822 (±.272)		-1.828** (±.241)	-2.876** (±.230)	-3.517** (±.310)	-3.918** (±.346)	-4.308** (±.386)	-4.526** (±.375)	-4.844** (±.426)
5	4.131** (±.503)	3.683** (±.353)	2.650** (±.335)	1.828** (±.241)		-1.048** (±.171)	-1.689** (±.248)	-2.090** (±.281)	-2.480** (±.335)	-2.698** (±.347)	-3.016** (±.409)
6	5.179** (±.424)	4.732** (±.295)	3.698** (±.276)	2.876** (±.230)	1.048** (±.171)		-.641 (±.203)	-1.042* (±.250)	-1.432** (±.291)	-1.650** (±.283)	-1.967** (±.337)
7	5.820** (±.444)	5.372** (±.357)	4.339** (±.321)	3.517** (±.310)	1.689** (±.248)	.641 (±.203)		-.401 (±.188)	-.791** (±.164)	-1.009** (±.178)	-1.327** (±.223)
8	6.221** (±.462)	5.773** (±.388)	4.740** (±.341)	3.918** (±.346)	2.090** (±.281)	1.042* (±.250)	.401 (±.188)		-.390 (±.134)	-.608* (±.162)	-.926** (±.199)
9	6.611** (±.471)	6.163** (±.401)	5.130** (±.380)	4.308** (±.386)	2.480** (±.335)	1.432** (±.291)	.791** (±.164)	.390 (±.134)		-.218 (±.124)	-.536* (±.136)
10	6.829** (±.443)	6.381** (±.394)	5.348** (±.369)	4.526** (±.375)	2.698** (±.347)	1.650** (±.283)	1.009** (±.178)	.608* (±.162)	.218 (±.124)		-.318 (±.109)
11	7.147** (±.468)	6.699** (±.420)	5.666** (±.395)	4.84** (±.426)	3.016** (±.409)	1.967** (±.337)	1.32** (±.223)	.926** (±.199)	.536* (±.136)	.318 (±.109)	

**P<.01 * P<.05

Pairwise comparison table depicting results based on the results of a two-way ANOVA depicting the relationship between Slope Error and the nSweeps condition for the newborn data at ten iterations.

SE	1	2	3	4	5	6	7	8	9	10	11
1		6.454 (±8.591)	27.431 (±10.573)	38.702 (±9.959)	72.537** (±8.938)	87.120** (±8.448)	93.919** (±9.392)	99.014** (±7.831)	108.758** (±8.605)	111.706** (±7.834)	114.699** (±8.615)
2	-6.454 (±8.591)		20.977 (±10.915)	32.248 (±9.157)	66.084** (±6.941)	80.666** (±9.158)	87.466** (±9.933)	92.560** (±8.832)	102.304** (±8.341)	105.252** (±7.995)	108.245** (±8.624)
3	-27.431 (±10.573)	-20.977 (±10.915)		11.271 (±10.964)	45.106** (±8.479)	59.689** (±8.634)	66.488** (±9.557)	71.583** (±9.899)	81.327** (±9.584)	84.275** (±9.514)	87.268** (±10.363)
4	-38.702* (±9.959)	-32.248 (±9.157)	-11.271 (±10.964)		33.836* (±7.957)	48.418** (±8.592)	55.218** (±9.365)	60.312** (±7.805)	70.056** (±9.129)	73.004** (±8.555)	75.997** (±8.920)
5	-72.537** (±8.938)	-66.084** (±6.941)	-45.106** (±8.479)	-33.836* (±7.957)		14.583 (±6.288)	21.382 (±7.941)	26.477 (±7.984)	36.221** (±8.091)	39.169** (±7.747)	42.162** (±8.645)
6	-87.120** (±8.448)	-80.666** (±9.158)	-59.689** (±8.634)	-48.418** (±8.592)	-14.583 (±6.288)		6.799 (±5.350)	11.894 (±6.150)	21.638* (±5.464)	24.586** (±5.479)	27.579* (±6.612)
7	-93.919** (±9.392)	-87.466** (±9.933)	-66.488** (±9.557)	-55.218** (±9.365)	-21.382 (±7.941)	-6.799 (±5.350)		5.095 (±4.956)	14.839 (±4.075)	17.787 (±4.949)	20.780** (±4.763)
8	-99.014** (±7.831)	-92.560** (±8.832)	-71.583** (±9.899)	-60.312** (±7.805)	-26.477 (±7.984)	-11.894 (±6.150)	-5.095 (±4.956)		9.744 (±4.289)	12.692 (±3.939)	15.685 (±4.547)
9	-108.758** (±8.605)	-102.304** (±8.341)	-81.327** (±9.584)	-70.056** (±9.129)	-36.221** (±8.091)	-21.638* (±5.464)	-14.839 (±4.075)	-9.744 (±4.289)		2.948 (±2.494)	5.941 (±2.899)
10	-111.706** (±7.834)	-105.252** (±7.995)	-84.275** (±9.514)	-73.004** (±8.555)	-39.169** (±7.747)	-24.586** (±5.479)	-17.787 (±4.949)	-12.692 (±3.939)	-2.948 (±2.494)		2.993 (±2.148)
11	-114.699** (±8.615)	-108.245** (±8.624)	-87.268** (±10.363)	-75.997** (±8.920)	-42.162** (±8.645)	-27.579* (±6.612)	-20.780** (±4.763)	-15.685 (±4.547)	-5.941 (±2.899)	-2.993 (±2.148)	

**P<.01 * P<.05

257

Pairwise comparison table depicting results based on the results of a two-way ANOVA depicting the relationship between Tracking Accuracy and the nSweeps condition for the newborn data at ten iterations.

TA	1	2	3	4	5	6	7	8	9	10	11
1		.020 (±.017)	.068 (±.020)	.106** (±.024)	.221** (±.030)	.275** (±.028)	.319** (±.026)	.350** (±.025)	.380** (±.026)	.392** (±.024)	.411** (±.026)
2	-.020 (±.017)		.048 (±.021)	.086* (±.021)	.201** (±.024)	.255** (±.027)	.299** (±.024)	.329** (±.026)	.360** (±.025)	.372** (±.025)	.391** (±.028)
3	-.068 (±.020)	-.048 (±.021)		.038 (±.019)	.153** (±.021)	.207** (±.022)	.251** (±.019)	.282** (±.021)	.312** (±.022)	.324** (±.023)	.343** (±.025)
4	-.106** (±.024)	-.086* (±.021)	-.038 (±.019)		.115** (±.020)	.169** (±.020)	.213** (±.020)	.243** (±.021)	.274** (±.023)	.286** (±.022)	.305** (±.024)
5	-.221** (±.030)	-.201** (±.024)	-.153** (±.021)	-.115** (±.020)		.054 (±.020)	.098** (±.022)	.128** (±.026)	.159** (±.027)	.171** (±.027)	.190** (±.031)
6	-.275** (±.028)	-.255** (±.027)	-.207** (±.022)	-.169** (±.020)	-.054 (±.020)		.044** (±.015)	.075* (±.018)	.105** (±.021)	.117** (±.020)	.136** (±.025)
7	-.319** (±.026)	-.299** (±.024)	-.251** (±.019)	-.213** (±.020)	-.098** (±.022)	-.044 (±.015)		.031 (±.017)	.061** (±.013)	.073** (±.016)	.092** (±.017)
8	-.350** (±.025)	-.329** (±.026)	-.282** (±.021)	-.243** (±.021)	-.128** (±.026)	-.075* (±.018)	-.031 (±.017)		.030 (±.014)	.043* (±.010)	.061* (±.015)
9	-.380** (±.026)	-.360** (±.025)	-.312** (±.022)	-.274** (±.023)	-.159** (±.027)	-.105** (±.021)	-.061** (±.013)	-.030 (±.014)		.012 (±.009)	.031 (±.010)
10	-.392** (±.024)	-.372** (±.025)	-.324** (±.023)	-.286** (±.022)	-.171** (±.027)	-.117** (±.020)	-.073** (±.016)	-.043* (±.010)	-.012 (±.009)		.019 (±.009)
11	-.411** (±.026)	-.391** (±.028)	-.343** (±.025)	-.305** (±.024)	-.190** (±.031)	-.136** (±.025)	-.092** (±.017)	-.061* (±.015)	-.031 (±.010)	-.019 (±.009)	

**P<.01 * P<.05

Appendix F

100 Iteration nSweeps Confusion Matrices

Pairwise comparison table depicting results based on the results of a one-way ANOVA depicting the relationship between Frequency Error and the nSweeps condition for the hidden component of newborn and adult data at one hundred iterations.

FE	1	2	3	4	5	6	7	8	9	10	11
1		.137 (±.163)	.756 (±.250)	1.727** (±.274)	2.917** (±.291)	3.662** (±.303)	4.084** (±.301)	4.414** (±.298)	4.653** (±.288)	4.803** (±.292)	4.973** (±.286)
2	-.137 (±.163)		.619** (±.134)	1.590** (±.196)	2.779** (±.260)	3.524** (±.290)	3.947** (±.292)	4.277** (±.292)	4.516** (±.287)	4.666** (±.293)	4.836** (±.287)
3	-.756 (±.250)	-.619** (±.134)		.971** (±.144)	2.160** (±.233)	2.905** (±.271)	3.327** (±.279)	3.657** (±.279)	3.896** (±.275)	4.047** (±.278)	4.217** (±.278)
4	-1. 727** (±.274)	-1.590** (±.196)	-.971** (±.144)		1.189** (±.133)	1.934** (±.183)	2.357** (±.199)	2.687** (±.209)	2.925** (±.210)	3.076** (±.221)	3.246** (±.224)
5	-2.917** (±.291)	-2.779** (±.260)	-2.160** (±.233)	-1.189** (±.133)		.745** (±.082)	1.167** (±.110)	1.497** (±.125)	1.736** (±.140)	1.887** (±.157)	2.056** (±.163)
6	-3.662** (±.303)	-3.524** (±.290)	-2.90** (±.271)	-1.934** (±.183)	-.745** (±.082)		.422** (±.051)	.752** (±.075)	.991** (±.094)	1.142** (±.116)	1.311** (±.127)
7	-4.084** (±.301)	-3.947** (±.292)	-3.327** (±.279)	-2.357** (±.199)	-1.167** (±.110)	-.422** (±.051)		.330** (±.044)	.569** (±.064)	.719** (±.090)	.889** (±.104)
8	-4.414** (±.298)	-4.277** (±.292)	-3.657** (±.279)	-2.687** (±.209)	-1.497** (±.125)	-.752** (±.075)	-.330** (±.044)		.239** (±.036)	.389** (±.059)	.559** (±.075)
9	-4.653** (±.288)	-4.516** (±.287)	-3.896** (±.275)	-2.925** (±.210)	-1.736** (±.140)	-.991** (±.094)	-.569** (±.064)	-.239** (±.036)		.151** (±.037)	.320** (±.053)
10	-4.803** (±.292)	-4.666** (±.293)	-4.047** (±.278)	-3.076** (±.221)	-1.887** (±.157)	-1.142** (±.116)	-.719** (±.090)	-.389** (±.059)	-.151** (±.037)		.170** (±.033)
11	-4.973** (±.286)	-4.836** (±.287)	-4.217** (±.278)	-3.246** (±.224)	-2.056** (±.163)	-1.311** (±.127)	-.889** (±.104)	-.559 (±.075)	-.320** (±.053)	-.170** (±.033)	

**P<.01 * P<.05

Pairwise comparison table depicting results based on the results of a one-way ANOVA depicting the relationship between Slope Error and the nSweeps condition for the hidden component of newborn and adult data at one hundred iterations.

SE	1	2	3	4	5	6	7	8	9	10	11
1		-17.772 (±2.102)	-36.506** (±3.172)	-57.823** (±4.204)	-76.677** (±4.732)	-89.071** (±5.180)	-94.588** (±5.274)	-99.561** (±5.356)	-102.446** (±5.330)	-104.793** (±5.445)	-107.350** (±5.405)
2	17.772** (±2.102)		-18.734** (±2.038)	-40.052** (±3.506)	-58.905** (±4.386)	-71.299** (±5.051)	-76.816** (±5.291)	-81.789** (±5.488)	-84.674** (±5.527)	-87.021** (±5.661)	-89.578** (±5.632)
3	36.506** (±3.172)	18.734** (±2.038)		-21.317** (±2.465)	-40.171** (±3.528)	-52.565** (±4.267)	-58.082** (±4.597)	-63.055** (±4.902)	-65.940** (±4.961)	-68.287** (±5.107)	-70.844** (±5.138)
4	57.823** (±4.204)	40.052** (±3.506)	21.317** (±2.465)		-18.854** (±1.851)	-31.247** (±2.681)	-36.765** (±3.164)	-41.737** (±3.615)	-44.623** (±3.756)	-46.969** (±4.032)	-49.527** (±4.134)
5	76.677** (±4.732)	58.905** (±4.386)	40.171** (±3.528)	18.854** (±1.851)		-12.394** (±1.441)	-17.911** (±2.091)	-22.884** (±2.593)	-25.769** (±2.873)	-28.116** (±3.239)	-30.673** (±3.407)
6	89.071** (±5.180)	71.299** (±5.051)	52.565** (±4.267)	31.247** (±2.681)	12.394** (±1.441)		-5.517** (±.987)	-10.490** (±1.570)	-13.376** (±1.966)	-15.722** (±2.370)	-18.279** (±2.579)
7	94.588** (±5.274)	76.816** (±5.291)	58.082** (±4.597)	36.765** (±3.164)	17.911** (±2.091)	5.517** (±.987)		-4.973** (±.845)	-7.858** (±1.184)	-10.205** (±1.608)	-12.762** (±1.850)
8	99.561** (±5.356)	81.789** (±5.488)	63.055** (±4.902)	41.737** (±3.615)	22.884** (±2.593)	10.490** (±1.570)	4.973** (±.845)		-2.885* (±.735)	-5.232** (±1.099)	-7.789** (±1.462)
9	102.446** (±5.330)	84.674** (±5.527)	65.940** (±4.961)	44.623** (±3.756)	25.769** (±2.873)	13.376** (±1.966)	7.858** (±1.184)	2.885* (±.735)		-2.347* (±.602)	-4.904** (±.926)
10	104.793** (±5.445)	87.021** (±5.661)	68.287** (±5.107)	46.969** (±4.032)	28.116** (±3.239)	15.722** (±2.370)	10.205** (±1.608)	5.232** (±1.099)	2.347* (±.602)		-2.557* (±.717)
11	107.350** (±5.405)	89.578** (±5.632)	70.844** (±5.138)	49.527** (±4.134)	30.673** (±3.407)	18.279** (±2.579)	12.762** (±1.850)	7.789** (±1.462)	4.904** (±.926)	2.557* (±.717)	

**P<.01 * P<.05

Pairwise comparison table depicting results based on the results of a one-way ANOVA depicting the relationship between Tracking Accuracy and the nSweeps condition for the hidden component of newborn and adult data at one hundred iterations.

TA	1	2	3	4	5	6	7	8	9	10	11
1		-.004 (±.010)	-.030 (±.015)	-.067* (±.018)	-.116** (±.019)	-.160** (±.020)	-.182** (±.019)	-.202** (±.019)	-.217** (±.018)	-.227** (±.019)	-.238** (±.018)
2	.004 (±.010)		-.026 (±.009)	-.063** (±.014)	-.112** (±.017)	-.156** (±.019)	-.178** (±.019)	-.198** (±.019)	-.213** (±.019)	-.223** (±.019)	-.234** (±.019)
3	.030 (±.015)	.026 (±.009)		-.037* (±.010)	-.086** (±.015)	-.130** (±.018)	-.152** (±.018)	-.173** (±.018)	-.187** (±.018)	-.197** (±.018)	-.208** (±.018)
4	.067* (±.018)	.063** (±.014)	.037* (±.010)		-.049** (±.009)	-.093** (±.012)	-.115** (±.013)	-.135** (±.014)	-.150** (±.014)	-.160** (±.015)	-.171** (±.015)
5	.116** (±.019)	.112** (±.017)	.086** (±.015)	.049** (±.009)		-.044** (±.006)	-.066** (±.008)	-.086** (±.009)	-.101** (±.010)	-.110** (±.012)	-.121** (±.012)
6	.160** (±.020)	.156** (±.019)	.130** (±.018)	.093** (±.012)	.044** (±.006)		-.022** (±.003)	-.042** (±.006)	-.057** (±.007)	-.066** (±.009)	-.077** (±.009)
7	.182** (±.019)	.178** (±.019)	.152** (±.018)	.115** (±.013)	.066** (±.008)	.022** (±.003)		-.020** (±.004)	-.035** (±.005)	-.045** (±.007)	-.056** (±.008)
8	.202** (±.019)	.198** (±.019)	.173** (±.018)	.135** (±.014)	.086** (±.009)	.042** (±.006)	.020** (±.004)		-.015** (±.002)	-.024** (±.005)	-.035** (±.006)
9	.217** (±.018)	.213** (±.019)	.187** (±.018)	.150** (±.014)	.101** (±.010)	.057** (±.007)	.035** (±.005)	.015** (±.002)		-.009 (±.003)	-.020** (±.004)
10	.227** (±.019)	.223** (±.019)	.197** (±.018)	.160** (±.015)	.110** (±.012)	.066** (±.009)	.045** (±.007)	.024** (±.005)	.009 (±.003)		-.011** (±.002)
11	.238** (±.018)	.234** (±.019)	.208** (±.018)	.171** (±.015)	.121** (±.012)	.077** (±.009)	.056** (±.008)	.035** (±.006)	.020** (±.004)	.011** (±.002)	

**P<.01 * P<.05

Pairwise comparison table depicting results based on the results of a one-way ANOVA depicting the relationship between Frequency Error and the nSweeps condition for the hidden component of the adult data at one hundred iterations.

FE	1	2	3	4	5	6	7	8	9	10	11
1		.566 (±.309)	1.576 (±.491)	2.826** (±.524)	4.150** (±.497)	4.650** (±.498)	4.923** (±.481)	5.079** (±.483)	5.198** (±.474)	5.268** (±.476)	5.364** (±.469)
2	-.566 (±.309)		1.010* (±.245)	2.260** (±.346)	3.584** (±.411)	4.084** (±.453)	4.356** (±.451)	4.513** (±.467)	4.632** (±.466)	4.701** (±.475)	4.798** (±.474)
3	-1.576 (±.491)	-1.010* (±.245)		1.250** (±.239)	2.574** (±.360)	3.074** (±.433)	3.346** (±.453)	3.503** (±.467)	3.622** (±.474)	3.691** (±.479)	3.788** (±.486)
4	-2.826** (±.524)	-2.260** (±.346)	-1.250** (±.239)		1.324** (±.191)	1.824** (±.283)	2.097** (±.309)	2.253** (±.336)	2.372** (±.346)	2.442** (±.359)	2.538** (±.370)
5	-4.150** (±.497)	-3.584** (±.411)	-2.574** (±.360)	-1.324** (±.191)		.501* (±.117)	.773** (±.160)	.929** (±.177)	1.048** (±.189)	1.118** (±.202)	1.214** (±.212)
6	-4.650** (±.498)	-4.084** (±.453)	-3.074** (±.433)	-1.824** (±.283)	-.501* (±.117)		.272* (±.069)	.428** (±.087)	.548** (±.099)	.617** (±.118)	.713** (±.134)
7	-4.923** (±.481)	-4.356** (±.451)	-3.346** (±.453)	-2.097** (±.309)	-.773** (±.160)	-.272* (±.069)		.156 (±.056)	.276* (±.065)	.345* (±.092)	.441* (±.106)
8	-5.079** (±.483)	-4.513** (±.467)	-3.503** (±.467)	-2.253** (±.336)	-.929** (±.177)	-.428** (±.087)	-.156* (±.056)		.119* (±.029)	.189* (±.047)	.285** (±.060)
9	-5.198** (±.474)	-4.632** (±.466)	-3.622** (±.474)	-2.372** (±.346)	-1.048** (±.189)	-.548** (±.099)	-.276* (±.065)	-.119* (±.029)		.069 (±.036)	.166 (±.047)
10	-5.268** (±.476)	-4.701** (±.475)	-3.691** (±.479)	-2.442** (±.359)	-1.118** (±.202)	-.617** (±.118)	-.345* (±.092)	-.189* (±.047)	-.069 (±.036)		.096 (±.027)
11	-5.364** (±.469)	-4.798** (±.474)	-3.788** (±.486)	-2.538** (±.370)	-1.214** (±.212)	-.713** (±.134)	-.441* (±.106)	-.285** (±.060)	-.166 (±.047)	-.096 (±.027)	

**P<.01 * P<.05

Pairwise comparison table depicting results based on the results of a one-way ANOVA depicting the relationship between Slope Error and the nSweeps condition for the hidden component of the adult data at one hundred iterations.

SE	1	2	3	4	5	6	7	8	9	10	11
1		-26.166** (±3.834)	-49.267** (±5.726)	-71.103** (±7.183)	-89.291** (±7.374)	-96.426** (±8.020)	-99.358** (±7.933)	-102.611** (±8.009)	-103.868** (±8.000)	-104.656** (±8.091)	-106.375** (±8.150)
2	26.166** (±3.834)		-23.101** (±3.342)	-44.937** (±5.450)	-63.126** (±6.479)	-70.260** (±7.520)	-73.192** (±7.629)	-76.445** (±7.824)	-77.703** (±7.970)	-78.490** (±8.077)	-80.209** (±8.232)
3	49.267** (±5.726)	23.101** (±3.342)		-21.836** (±4.136)	-40.025** (±5.607)	-47.159** (±6.861)	-50.091** (±7.057)	-53.344** (±7.403)	-54.602** (±7.492)	-55.389** (±7.620)	-57.108** (±7.809)
4	71.103** (±7.183)	44.937** (±5.450)	21.836** (±4.136)		-18.188** (±2.675)	-25.323** (±3.960)	-28.255** (±4.368)	-31.508** (±4.915)	-32.765** (±5.121)	-33.553** (±5.440)	-35.272** (±5.715)
5	89.291** (±7.374)	63.126** (±6.479)	40.025** (±5.607)	18.188** (±2.675)		-7.134* (±1.798)	-10.066* (±2.394)	-13.320** (±2.761)	-14.577** (±3.150)	-15.365** (±3.488)	-17.084** (±3.889)
6	96.426** (±8.020)	70.260** (±7.520)	47.159** (±6.861)	25.323** (±3.960)	7.134* (±1.798)		-2.932 (±1.126)	-6.186* (±1.580)	-7.443 (±2.154)	-8.231 (±2.565)	-9.950 (±3.018)
7	99.358** (±7.933)	73.192** (±7.629)	50.091** (±7.057)	28.255** (±4.368)	10.066* (±2.394)	2.932 (±1.126)		-3.253 (±.995)	-4.511 (±1.343)	-5.298 (±1.776)	-7.018 (±2.217)
8	102.611** (±8.009)	76.445** (±7.824)	53.344** (±7.403)	31.508** (±4.915)	13.320** (±2.761)	6.186* (±1.580)	3.253 (±.995)		-1.257 (±.934)	-2.045 (±1.255)	-3.764 (±1.782)
9	103.868** (±8.000)	77.703** (±7.970)	54.602** (±7.492)	32.765** (±5.121)	14.577** (±3.150)	7.443 (±2.154)	4.511 (±1.343)	1.257 (±.934)		-.788 (±.622)	-2.507 (±1.085)
10	104.656** (±8.091)	78.490** (±8.077)	55.389** (±7.620)	33.553** (±5.440)	15.365** (±3.488)	8.231 (±2.565)	5.298 (±1.776)	2.045 (±1.255)	.788 (±.622)		-1.719 (±.916)
11	106.375** (±8.150)	80.209** (±8.232)	57.108** (±7.809)	35.272** (±5.715)	17.084** (±3.889)	9.950 (±3.018)	7.018 (±2.217)	3.764 (±1.782)	2.507 (±1.085)	1.719 (±.916)	

**P<.01 * P<.05

Pairwise comparison table depicting results based on the results of a one-way ANOVA depicting the relationship between Tracking Accuracy and the nSweeps condition for the hidden component of the adult data at one hundred iterations.

TA	1	2	3	4	5	6	7	8	9	10	11
1		-.029 (±.018)	-.085 (±.029)	-.148** (±.032)	-.218** (±.030)	-.246** (±.031)	-.258** (±.030)	-.269** (±.029)	-.275** (±.029)	-.276** (±.030)	-.282** (±.030)
2	.029 (±.018)		-.056 (±.016)	-.119** (±.024)	-.190** (±.025)	-.217** (±.028)	-.229** (±.028)	-.240** (±.029)	-.247** (±.028)	-.248** (±.030)	-.253** (±.030)
3	.085 (±.029)	.056 (±.016)		-.063* (±.015)	-.134** (±.021)	-.161** (±.026)	-.173** (±.027)	-.184** (±.028)	-.191** (±.028)	-.192** (±.029)	-.197** (±.029)
4	.148** (±.032)	.119** (±.024)	.063* (±.015)		-.071** (±.011)	-.098** (±.017)	-.110** (±.019)	-.121** (±.021)	-.128** (±.021)	-.129** (±.023)	-.134** (±.024)
5	.218** (±.030)	.190** (±.025)	.134** (±.021)	.071** (±.011)		-.028 (±.008)	-.039* (±.010)	-.050** (±.011)	-.057** (±.012)	-.058* (±.014)	-.064* (±.015)
6	.246** (±.031)	.217** (±.028)	.161** (±.026)	.098** (±.017)	.028 (±.008)		-.012 (±.004)	-.023* (±.006)	-.030* (±.007)	-.031 (±.010)	-.036 (±.011)
7	.258** (±.030)	.229** (±.028)	.173** (±.027)	.110** (±.019)	.039* (±.010)	.012 (±.004)		-.011 (±.005)	-.018 (±.006)	-.019 (±.009)	-.024 (±.011)
8	.269** (±.029)	.240** (±.029)	.184** (±.028)	.121** (±.021)	.050** (±.011)	.023* (±.006)	.011 (±.005)		-.007 (±.002)	-.008 (±.005)	-.013 (±.007)
9	.275** (±.029)	.247** (±.028)	.191** (±.028)	.128** (±.021)	.057** (±.012)	.030* (±.007)	.018 (±.006)	.007 (±.002)		-.001 (±.004)	-.006 (±.005)
10	.276** (±.030)	.248** (±.030)	.192** (±.029)	.129** (±.023)	.058* (±.014)	.031 (±.010)	.019 (±.009)	.008 (±.005)	.001 (±.004)		-.005 (±.003)
11	.282** (±.030)	.253** (±.030)	.197** (±.029)	.134** (±.024)	.064* (±.015)	.036 (±.011)	.024 (±.011)	.013 (±.007)	.006 (±.005)	.005 (±.003)	

**P<.01 * P<.05

Pairwise comparison table depicting results based on the results of a one-way ANOVA depicting the relationship between Frequency Error and the nSweeps condition for the hidden component of the newborn data at one hundred iterations.

FE	1	2	3	4	5	6	7	8	9	10	11
1		-.292 (±.116)	-.063 (±.131)	.628 (±.184)	1.683** (±.311)	2.673** (±.351)	3.245** (±.366)	3.749** (±.354)	4.107** (±.333)	4.339** (±.344)	4.582** (±.332)
2	.292 (±.116)		.229 (±.117)	.921** (±.194)	1.975** (±.322)	2.965** (±.364)	3.537** (±.375)	4.041** (±.356)	4.399** (±.340)	4.631** (±.347)	4.874** (±.331)
3	.063 (±.131)	-.229 (±.117)		.692* (±.165)	1.747** (±.298)	2.736** (±.329)	3.308** (±.332)	3.812** (±.311)	4.170** (±.288)	4.402** (±.289)	4.645** (±.281)
4	-.628 (±.184)	-.921** (±.194)	-.692* (±.165)		1.055** (±.185)	2.044** (±.232)	2.617** (±.253)	3.121** (±.251)	3.479** (±.242)	3.711** (±.261)	3.954** (±.257)
5	-1.683** (±.311)	-1.975** (±.322)	-1.747** (±.298)	-1.055** (±.185)		.989** (±.115)	1.562** (±.152)	2.066** (±.177)	2.424** (±.206)	2.656** (±.240)	2.899** (±.246)
6	-2.673** (±.351)	-2.965** (±.364)	-2.736** (±.329)	-2.044** (±.232)	-.989** (±.115)		.573** (±.075)	1.076** (±.121)	1.435** (±.158)	1.666** (±.197)	1.910** (±.214)
7	-3.245** (±.366)	-3.537** (±.375)	-3.308** (±.332)	-2.617** (±.253)	-1.562** (±.152)	-.573** (±.075)		.504** (±.067)	.862** (±.109)	1.094** (±.153)	1.337** (±.176)
8	-3.749** (±.354)	-4.041** (±.356)	-3.812** (±.311)	-3.121** (±.251)	-2.066** (±.177)	-1.076** (±.121)	-.504** (±.067)		.358** (±.065)	.590** (±.107)	.833** (±.136)
9	-4.107** (±.333)	-4.399** (±.340)	-4.170** (±.288)	-3.479** (±.242)	-2.424** (±.206)	-1.435** (±.158)	-.862** (±.109)	-.358** (±.065)		.232 (±.063)	.475** (±.094)
10	-4.339** (±.344)	-4.631** (±.347)	-4.402** (±.289)	-3.711** (±.261)	-2.656** (±.240)	-1.666** (±.197)	-1.094** (±.153)	-.590** (±.107)	-.232 (±.063)		.243* (±.059)
11	-4.582** (±.332)	-4.874** (±.331)	-4.645** (±.281)	-3.954** (±.257)	-2.899** (±.246)	-1.910** (±.214)	-1.337** (±.176)	-.833** (±.136)	-.475** (±.094)	-.243* (±.059)	

**P<.01 * P<.05

Pairwise comparison table depicting results based on the results of a one-way ANOVA depicting the relationship between Slope Error and the nSweeps condition for the hidden component of the newborn data at one hundred iterations.

SE	1	2	3	4	5	6	7	8	9	10	11
1		-9.378** ($\pm$1.833)	-23.745** ($\pm$2.881)	-44.544** ($\pm$4.493)	-64.062** ($\pm$5.982)	-81.716** ($\pm$6.609)	-89.818** ($\pm$6.982)	-96.510** ($\pm$7.142)	-101.024** ($\pm$7.074)	-104.930** ($\pm$7.312)	-108.324** ($\pm$7.134)
2	9.378** ($\pm$1.833)		-14.367** ($\pm$2.373)	-35.166** ($\pm$4.448)	-54.685** ($\pm$5.931)	-72.338** ($\pm$6.768)	-80.440** ($\pm$7.338)	-87.133** ($\pm$7.698)	-91.646** ($\pm$7.667)	-95.552** ($\pm$7.933)	-98.947** ($\pm$7.703)
3	23.745** ($\pm$2.881)	14.367** ($\pm$2.373)		-20.799** ($\pm$2.744)	-40.317** ($\pm$4.332)	-57.970** ($\pm$5.144)	-66.073** ($\pm$5.933)	-72.765** ($\pm$6.461)	-77.279** ($\pm$6.536)	-81.184** ($\pm$6.826)	-84.579** ($\pm$6.716)
4	44.544** ($\pm$4.493)	35.166** ($\pm$4.448)	20.799** ($\pm$2.744)		-19.519** ($\pm$2.562)	-37.172** ($\pm$3.626)	-45.274** ($\pm$4.570)	-51.967** ($\pm$5.286)	-56.480** ($\pm$5.480)	-60.386** ($\pm$5.932)	-63.781** ($\pm$5.962)
5	64.062** ($\pm$5.982)	54.685** ($\pm$5.931)	40.317** ($\pm$4.332)	19.519** ($\pm$2.562)		-17.653** ($\pm$2.238)	-25.756** ($\pm$3.396)	-32.448** ($\pm$4.342)	-36.961** ($\pm$4.757)	-40.867** ($\pm$5.401)	-44.262** ($\pm$5.542)
6	81.716** ($\pm$6.609)	72.338** ($\pm$6.768)	57.970** ($\pm$5.144)	37.172** ($\pm$3.626)	17.653** ($\pm$2.238)		-8.103** ($\pm$1.606)	-14.795** ($\pm$2.681)	-19.308** ($\pm$3.256)	-23.214** ($\pm$3.945)	-26.609** ($\pm$4.145)
7	89.818** ($\pm$6.982)	80.440** ($\pm$7.338)	66.073** ($\pm$5.933)	45.274** ($\pm$4.570)	25.756** ($\pm$3.396)	8.103** ($\pm$1.606)		-6.692** ($\pm$1.354)	-11.206** ($\pm$1.933)	-15.111** ($\pm$2.654)	-18.506** ($\pm$2.938)
8	96.510** ($\pm$7.142)	87.133** ($\pm$7.698)	72.765** ($\pm$6.461)	51.967** ($\pm$5.286)	32.448** ($\pm$4.342)	14.795** ($\pm$2.681)	6.692** ($\pm$1.354)		-4.513* ($\pm$1.128)	-8.419** ($\pm$1.787)	-11.814** ($\pm$2.301)
9	101.024** ($\pm$7.074)	91.646** ($\pm$7.667)	77.279** ($\pm$6.536)	56.480** ($\pm$5.480)	36.961** ($\pm$4.757)	19.308** ($\pm$3.256)	11.206** ($\pm$1.933)	4.513* ($\pm$1.128)		-3.906* ($\pm$1.019)	-7.301** ($\pm$1.487)
10	104.930** ($\pm$7.312)	95.552** ($\pm$7.933)	81.184** ($\pm$6.826)	60.386** ($\pm$5.932)	40.867** ($\pm$5.401)	23.214** ($\pm$3.945)	15.111** ($\pm$2.654)	8.419** ($\pm$1.787)	3.906* ($\pm$1.019)		-3.395 ($\pm$1.097)
11	108.324** ($\pm$7.134)	98.947** ($\pm$7.703)	84.579** ($\pm$6.716)	63.781** ($\pm$5.962)	44.262** ($\pm$5.542)	26.60** ($\pm$4.145)	18.506** ($\pm$2.938)	11.814** ($\pm$2.301)	7.301** ($\pm$1.487)	3.395 ($\pm$1.097)	

**P<.01 * P<.05

Pairwise comparison table depicting results based on the results of a one-way ANOVA depicting the relationship between Tracking Accuracy and the nSweeps condition for the hidden component of the newborn data at one hundred iterations.

TA	1	2	3	4	5	6	7	8	9	10	11
1		.021 ($\pm$.011)	.025 ($\pm$.011)	.014 ($\pm$.016)	-.014 ($\pm$.023)	-.075 ($\pm$.025)	-.106** ($\pm$.024)	-.136** ($\pm$.024)	-.159** ($\pm$.023)	-.177** ($\pm$.023)	-.193** ($\pm$.022)
2	-.021 ($\pm$.011)		.004 ($\pm$.008)	-.007 ($\pm$.015)	-.035 ($\pm$.023)	-.095 ($\pm$.026)	-.127** ($\pm$.026)	-.157** ($\pm$.026)	-.180** ($\pm$.025)	-.198** ($\pm$.025)	-.214** ($\pm$.023)
3	-.025 ($\pm$.011)	-.004 ($\pm$.008)		-.011 ($\pm$.013)	-.039 ($\pm$.023)	-.100* ($\pm$.025)	-.131** ($\pm$.025)	-.161** ($\pm$.024)	-.184** ($\pm$.022)	-.202** ($\pm$.022)	-.218** ($\pm$.021)
4	-.014 ($\pm$.016)	.007 ($\pm$.015)	.011 ($\pm$.013)		-.028 ($\pm$.014)	-.088** ($\pm$.016)	-.120** ($\pm$.017)	-.150** ($\pm$.018)	-.173** ($\pm$.017)	-.191** ($\pm$.019)	-.207** ($\pm$.017)
5	.014 ($\pm$.023)	.035 ($\pm$.023)	.039 ($\pm$.023)	.028 ($\pm$.014)		-.060** ($\pm$.010)	-.092** ($\pm$.011)	-.122** ($\pm$.015)	-.145** ($\pm$.016)	-.163** ($\pm$.019)	-.179** ($\pm$.018)
6	.075 ($\pm$.025)	.095 ($\pm$.026)	.100* ($\pm$.025)	.088** ($\pm$.016)	.060** ($\pm$.010)		-.032** ($\pm$.005)	-.062** ($\pm$.010)	-.084** ($\pm$.011)	-.102** ($\pm$.015)	-.119** ($\pm$.015)
7	.106** ($\pm$.024)	.127** ($\pm$.026)	.131** ($\pm$.025)	.120** ($\pm$.017)	.092** ($\pm$.011)	.032** ($\pm$.005)		-.030** ($\pm$.006)	-.053** ($\pm$.008)	-.070** ($\pm$.012)	-.087** ($\pm$.012)
8	.136** ($\pm$.024)	.157** ($\pm$.026)	.161** ($\pm$.024)	.150** ($\pm$.018)	.122** ($\pm$.015)	.062** ($\pm$.010)	.030** ($\pm$.006)		-.023** ($\pm$.004)	-.040** ($\pm$.009)	-.057** ($\pm$.009)
9	.159** ($\pm$.023)	.180** ($\pm$.025)	.184** ($\pm$.022)	.173** ($\pm$.017)	.145** ($\pm$.016)	.084** ($\pm$.011)	.053** ($\pm$.008)	.023** ($\pm$.004)		-.018 ($\pm$.006)	-.034** ($\pm$.006)
10	.177** ($\pm$.023)	.198** ($\pm$.025)	.202** ($\pm$.022)	.191** ($\pm$.019)	.163** ($\pm$.019)	.102** ($\pm$.015)	.070** ($\pm$.012)	.040** ($\pm$.009)	.018 ($\pm$.006)		-.016* ($\pm$.004)
11	.193** ($\pm$.022)	.214** ($\pm$.023)	.218** ($\pm$.021)	.207** ($\pm$.017)	.179** ($\pm$.018)	.119** ($\pm$.015)	.087** ($\pm$.012)	.057** ($\pm$.009)	.034** ($\pm$.006)	.016* ($\pm$.004)	

**P<.01 * P<.05

Pairwise comparison table depicting results based on the results of a two-way ANOVA depicting the relationship between Frequency Error and the nSweeps condition for the newborn and adult data at one hundred iterations.

FE	1	2	3	4	5	6	7	8	9	10	11
1		-1.105** (±.140)	-2.273** (±.190)	-3.711** (±.231)	-5.111** (±.255)	-5.868** (±.267)	-6.409** (±.268)	-6.668** (±.274)	-6.931** (±.277)	-7.160** (±.279)	-7.346** (±.290)
2	1.105** (±.140)		-1.168** (±.086)	-2.606** (±.165)	-4.006** (±.216)	-4.763** (±.246)	-5.305** (±.255)	-5.563** (±.266)	-5.826** (±.273)	-6.055** (±.281)	-6.242** (±.294)
3	2.273** (±.190)	1.168** (±.086)		-1.438** (±.108)	-2.838** (±.171)	-3.595** (±.209)	-4.136** (±.229)	-4.395** (±.245)	-4.658** (±.257)	-4.887** (±.269)	-5.073** (±.286)
4	3.711** (±.231)	2.606** (±.165)	1.438** (±.108)		-1.400** (±.096)	-2.157** (±.143)	-2.698** (±.179)	-2.957** (±.202)	-3.220** (±.221)	-3.449** (±.240)	-3.635** (±.262)
5	5.111** (±.255)	4.006** (±.216)	2.838** (±.171)	1.400** (±.096)		-.757** (±.063)	-1.299** (±.102)	-1.558** (±.130)	-1.821** (±.152)	-2.050** (±.175)	-2.236** (±.199)
6	5.868** (±.267)	4.763** (±.246)	3.595** (±.209)	2.157** (±.143)	.757** (±.063)		-.542** (±.059)	-.800** (±.086)	-1.063** (±.110)	-1.292** (±.134)	-1.479** (±.162)
7	6.409** (±.268)	5.305** (±.255)	4.136** (±.229)	2.698** (±.179)	1.299** (±.102)	.542** (±.059)		-.259** (±.045)	-.522** (±.061)	-.751** (±.087)	-.937** (±.111)
8	6.668** (±.274)	5.563** (±.266)	4.395** (±.245)	2.957** (±.202)	1.558** (±.130)	.800** (±.086)	.259** (±.045)		-.263** (±.040)	-.492** (±.063)	-.678** (±.090)
9	6.931** (±.277)	5.826** (±.273)	4.658** (±.257)	3.220** (±.221)	1.821** (±.152)	1.063** (±.110)	.522** (±.061)	.263** (±.040)		-.229** (±.037)	-.415** (±.064)
10	7.160** (±.290)	6.055** (±.281)	4.887** (±.269)	3.449** (±.240)	2.050** (±.175)	1.292** (±.134)	.751** (±.087)	.492** (±.063)	.229** (±.037)		-.186** (±.039)
11	7.346** (±.279)	6.242** (±.294)	5.073** (±.286)	3.635** (±.262)	2.236** (±.199)	1.479** (±.162)	.937** (±.111)	.678** (±.090)	.415** (±.064)	.186** (±.039)	

**P<.01 * P<.05

Pairwise comparison table depicting results based on the results of a two-way ANOVA depicting the relationship between Slope Error and the nSweeps condition for the newborn and adult data at one hundred iterations.

SE	1	2	3	4	5	6	7	8	9	10	11
1		23.396** (±2.823)	41.286** (±3.344)	63.808** (±3.947)	81.344** (±4.463)	91.682** (±4.807)	98.987** (±4.963)	102.257** (±5.300)	106.105** (±5.554)	108.472** (±5.852)	111.135** (±6.116)
2	-23.396** (±2.823)		17.890** (±1.935)	40.412** (±2.710)	57.948** (±3.457)	68.286** (±4.167)	75.591** (±4.394)	78.861** (±4.815)	82.709** (±5.144)	85.076** (±5.454)	87.739** (±5.742)
3	-41.286** (±3.344)	-17.890** (±1.935)		22.522** (±2.233)	40.058** (±3.123)	50.396** (±4.054)	57.701** (±4.364)	60.971** (±4.816)	64.818** (±5.217)	67.186** (±5.519)	69.849** (±5.822)
4	-63.808** (±3.947)	-40.412** (±2.710)	-22.522** (±2.233)		17.536** (±2.078)	27.874** (±2.793)	35.179** (±3.374)	38.449** (±3.932)	42.297** (±4.427)	44.664** (±4.869)	47.327** (±5.204)
5	-81.344** (±4.463)	-57.948** (±3.457)	-40.058** (±3.123)	-17.536** (±2.078)		10.339** (±1.708)	17.643** (±1.977)	20.913** (±2.524)	24.761** (±3.013)	27.128** (±3.444)	29.791** (±3.811)
6	-91.682** (±4.807)	-68.286** (±4.167)	-50.396** (±4.054)	-27.874** (±2.793)	-10.339** (±1.708)		7.305** (±1.202)	10.575** (±1.695)	14.422** (±2.223)	16.790** (±2.782)	19.453** (±3.176)
7	-98.987** (±4.963)	-75.591** (±4.394)	-57.701** (±4.364)	-35.179** (±3.374)	-17.643** (±1.977)	-7.305** (±1.202)		3.270 (±.972)	7.118** (±1.420)	9.485** (±1.987)	12.148** (±2.356)
8	-102.257** (±5.300)	-78.861** (±4.815)	-60.971** (±4.816)	-38.449** (±3.932)	-20.913** (±2.524)	-10.575** (±1.695)	-3.270 (±.972)		3.848** (±.949)	6.215** (±1.511)	8.878** (±1.900)
9	-106.105** (±5.554)	-82.709** (±5.144)	-64.818** (±5.217)	-42.297** (±4.427)	-24.761** (±3.013)	-14.422** (±2.223)	-7.118** (±1.420)	-3.848** (±.949)		2.367 (±.849)	5.031** (±1.203)
10	-108.472** (±5.852)	-85.076** (±5.454)	-67.186** (±5.519)	-44.664** (±4.869)	-27.128** (±3.444)	-16.790** (±2.782)	-9.485** (±1.987)	-6.215** (±1.511)	-2.367 (±.849)		2.663** (±.628)
11	-111.135** (±6.116)	-87.739** (±5.742)	-69.849** (±5.822)	-47.327** (±5.204)	-29.791** (±3.811)	-19.453** (±3.176)	-12.148** (±2.356)	-8.878** (±1.900)	-5.031** (±1.203)	-2.663** (±.628)	

**P<.01 * P<.05

Pairwise comparison table depicting results based on the results of a two-way ANOVA depicting the relationship between Tracking Accuracy and the nSweeps condition for the newborn and adult data at one hundred iterations.

TA	1	2	3	4	5	6	7	8	9	10	11
1		.061** (±.007)	.118** (±.011)	.189** (±.014)	.260** (±.015)	.302** (±.016)	.334** (±.016)	.349** (±.017)	.368** (±.017)	.382** (±.017)	.395** (±.017)
2	-.061** (±.007)		.058** (±.005)	.129** (±.009)	.199** (±.012)	.242** (±.014)	.273** (±.015)	.289** (±.016)	.307** (±.016)	.321** (±.017)	.335** (±.017)
3	-.118** (±.011)	-.058** (±.005)		.071** (±.007)	.142** (±.010)	.184** (±.013)	.216** (±.014)	.231** (±.015)	.250** (±.016)	.264** (±.017)	.277** (±.017)
4	-.189** (±.014)	-.129** (±.009)	-.071** (±.007)		.071** (±.006)	.113** (±.009)	.145** (±.011)	.160** (±.013)	.178** (±.014)	.192** (±.015)	.206** (±.016)
5	-.260** (±.015)	-.199** (±.012)	-.142** (±.010)	-.071** (±.006)		.042** (±.005)	.074** (±.007)	.089** (±.009)	.108** (±.010)	.122** (±.012)	.135** (±.013)
6	-.302** (±.016)	-.242** (±.014)	-.184** (±.013)	-.113** (±.009)	-.042** (±.005)		.032** (±.004)	.047** (±.006)	.065** (±.007)	.079** (±.009)	.093** (±.010)
7	-.334** (±.016)	-.273** (±.015)	-.216** (±.014)	-.145** (±.011)	-.074** (±.007)	-.032** (±.004)		.015** (±.003)	.034** (±.004)	.048** (±.006)	.061** (±.008)
8	-.349** (±.017)	-.289** (±.016)	-.231** (±.015)	-.160** (±.013)	-.089** (±.009)	-.047** (±.006)	-.015** (±.003)		.019** (±.003)	.032** (±.005)	.046** (±.006)
9	-.368** (±.017)	-.307** (±.016)	-.250** (±.016)	-.178** (±.014)	-.108** (±.010)	-.065** (±.007)	-.034** (±.004)	-.019** (±.003)		.014** (±.003)	.028** (±.004)
10	-.382** (±.017)	-.321** (±.017)	-.264** (±.017)	-.192** (±.015)	-.122** (±.012)	-.079** (±.009)	-.048** (±.006)	-.032** (±.005)	-.014** (±.003)		.014** (±.003)
11	-.395** (±.017)	-.335** (±.017)	-.277** (±.017)	-.206** (±.016)	-.135** (±.013)	-.093** (±.010)	-.061** (±.008)	-.046** (±.006)	-.028** (±.004)	-.014** (±.003)	

**P<.01 * P<.05

Pairwise comparison table depicting results based on the results of a two-way ANOVA depicting the relationship between Frequency Error and the nSweeps condition for the adult data at one hundred iterations.

FE	1	2	3	4	5	6	7	8	9	10	11
1		-1.847** (±.248)	-3.354** (±.336)	-4.966** (±.371)	-6.183** (±.391)	-6.727** (±.417)	-7.072** (±.407)	-7.225** (±.422)	-7.414** (±.411)	-7.538** (±.417)	-7.634** (±.413)
2	1.847** (±.248)		-1.507** (±.135)	-3.119** (±.254)	-4.336** (±.337)	-4.880** (±.395)	-5.225** (±.401)	-5.378** (±.425)	-5.567** (±.422)	-5.691** (±.434)	-5.787** (±.437)
3	3.354** (±.336)	1.507** (±.135)		-1.612** (±.171)	-2.829** (±.284)	-3.373** (±.351)	-3.718** (±.371)	-3.871** (±.397)	-4.060** (±.402)	-4.184** (±.417)	-4.280** (±.427)
4	4.966** (±.371)	3.119** (±.254)	1.612** (±.171)		-1.217** (±.150)	-1.761** (±.223)	-2.106** (±.257)	-2.259** (±.290)	-2.448** (±.301)	-2.572** (±.326)	-2.668** (±.341)
5	6.183** (±.391)	4.336** (±.337)	2.829** (±.284)	1.217** (±.150)		-.544** (±.087)	-.889** (±.120)	-1.042** (±.153)	-1.231** (±.169)	-1.355** (±.196)	-1.450** (±.216)
6	6.727** (±.417)	4.880** (±.395)	3.373** (±.351)	1.761** (±.223)	.544** (±.087)		-.345** (±.061)	-.498** (±.093)	-.687** (±.112)	-.811** (±.141)	-.907** (±.165)
7	7.072** (±.407)	5.225** (±.401)	3.718** (±.371)	2.106** (±.257)	.889** (±.120)	.345** (±.061)		-.153 (±.050)	-.342** (±.060)	-.466** (±.090)	-.561** (±.110)
8	7.225** (±.422)	5.378** (±.425)	3.871** (±.397)	2.259** (±.290)	1.042** (±.153)	.498** (±.093)	.153 (±.050)		-.189 (±.053)	-.313** (±.070)	-.409** (±.088)
9	7.414** (±.411)	5.567** (±.422)	4.060** (±.402)	2.448** (±.301)	1.231** (±.169)	.687** (±.112)	.342** (±.060)	.189 (±.053)		-.124 (±.039)	-.220 (±.060)
10	7.538** (±.417)	5.691** (±.434)	4.184** (±.417)	2.572** (±.326)	1.355** (±.196)	.811** (±.141)	.466** (±.090)	.313** (±.070)	.124 (±.039)		-.096 (±.040)
11	7.634** (±.413)	5.787** (±.437)	4.280** (±.427)	2.668** (±.341)	1.450** (±.216)	.907** (±.165)	.561** (±.110)	.409** (±.088)	.220 (±.060)	.096 (±.040)	

**P<.01 * P<.05

Pairwise comparison table depicting results based on the results of a two-way ANOVA depicting the relationship between Slope Error and the nSweeps condition for the adult data at one hundred iterations.

SE	1	2	3	4	5	6	7	8	9	10	11
1		34.489** (±3.511)	55.507** (±4.648)	78.008** (±5.727)	89.510** (±7.097)	97.276** (±7.643)	101.045** (±7.924)	101.985** (±8.669)	105.242** (±9.038)	106.205** (±9.562)	106.985** (±9.870)
2	-34.489** (±3.511)		21.018** (±2.931)	43.519** (±4.331)	55.021** (±5.920)	62.787** (±6.741)	66.556** (±7.198)	67.496** (±7.842)	70.753** (±8.300)	71.716** (±8.859)	72.496** (±9.170)
3	-55.507** (±4.648)	-21.018** (±2.931)		22.501** (±3.785)	34.003** (±5.482)	41.769** (±6.599)	45.538** (±7.170)	46.478** (±7.861)	49.735** (±8.366)	50.698** (±8.873)	51.479** (±9.266)
4	-78.008** (±5.727)	-43.519** (±4.331)	-22.501** (±3.785)		11.502 (±3.246)	19.268** (±4.223)	23.037** (±4.943)	23.978* (±5.852)	27.234* (±6.562)	28.197* (±7.345)	28.978* (±7.726)
5	-89.510** (±7.097)	-55.021** (±5.920)	-34.003** (±5.482)	-11.502 (±3.246)		7.766* (±1.869)	11.535** (±2.463)	12.475* (±3.209)	15.732* (±3.985)	16.695 (±4.683)	17.476 (±5.105)
6	-97.276** (±7.643)	-62.787** (±6.741)	-41.769** (±6.599)	-19.268** (±4.223)	-7.766* (±1.869)		3.769 (±1.037)	4.710 (±1.854)	7.966 (±2.861)	8.929 (±3.750)	9.710 (±4.122)
7	-101.045** (±7.924)	-66.556** (±7.198)	-45.538** (±7.170)	-23.037** (±4.943)	-11.535** (±2.463)	-3.769 (±1.037)		.941 (±1.375)	4.197 (±2.189)	5.160 (±3.070)	5.941 (±3.457)
8	-101.985** (±8.669)	-67.496** (±7.842)	-46.478** (±7.861)	-23.978* (±5.852)	-12.475* (±3.209)	-4.710 (±1.854)	-.941 (±1.375)		3.256 (±1.369)	4.219 (±2.142)	5.000 (±2.464)
9	-105.242** (±9.038)	-70.753** (±8.300)	-49.735** (±8.366)	-27.234* (±6.562)	-15.732* (±3.985)	-7.966 (±2.861)	-4.197 (±2.189)	-3.256 (±1.369)		.963 (±1.054)	1.744 (±1.385)
10	-106.205** (±9.562)	-71.716** (±8.859)	-50.698** (±8.873)	-28.197* (±7.345)	-16.695 (±4.683)	-8.929 (±3.750)	-5.160 (±3.070)	-4.219 (±2.142)	-.963 (±1.054)		.781 (±.587)
11	-106.985** (±9.870)	-72.496** (±9.170)	-51.479** (±9.266)	-28.978* (±7.726)	-17.476 (±5.105)	-9.710 (±4.122)	-5.941 (±3.457)	-5.000 (±2.464)	-1.744 (±1.385)	-.781 (±.587)	

**P<.01 * P<.05

Pairwise comparison table depicting results based on the results of a two-way ANOVA depicting the relationship between Tracking Accuracy and the nSweeps condition for the adult data at one hundred iterations.

TA	1	2	3	4	5	6	7	8	9	10	11
1		.098** (±.012)	.172** (±.018)	.251** (±.021)	.307** (±.023)	.335** (±.025)	.350** (±.025)	.357** (±.026)	.371* (±.026)	.377** (±.026)	.383** (±.025)
2	-.098** (±.012)		.074** (±.009)	.152** (±.015)	.209** (±.020)	.237** (±.023)	.252** (±.024)	.259** (±.025)	.272** (±.026)	.279** (±.026)	.285** (±.026)
3	-.172** (±.018)	-.074** (±.009)		.078** (±.011)	.135** (±.017)	.163** (±.022)	.178** (±.023)	.185** (±.025)	.199** (±.026)	.205** (±.026)	.211** (±.026)
4	-.251** (±.021)	-.152** (±.015)	-.078** (±.011)		.057** (±.009)	.084** (±.014)	.100** (±.016)	.107** (±.018)	.120** (±.020)	.126** (±.021)	.132** (±.021)
5	-.307** (±.023)	-.209** (±.020)	-.135** (±.017)	-.057** (±.009)		.028** (±.006)	.043** (±.009)	.050** (±.011)	.063** (±.013)	.070** (±.014)	.076** (±.015)
6	-.335** (±.025)	-.237** (±.023)	-.163** (±.022)	-.084** (±.014)	-.028** (±.006)		.016** (±.003)	.023* (±.005)	.036** (±.008)	.042** (±.009)	.048** (±.010)
7	-.350** (±.025)	-.252** (±.024)	-.178** (±.023)	-.100** (±.016)	-.043** (±.009)	-.016** (±.003)		.007 (±.003)	.020** (±.005)	.027* (±.006)	.033** (±.007)
8	-.357** (±.026)	-.259** (±.025)	-.185** (±.025)	-.107** (±.018)	-.050** (±.011)	-.023* (±.005)	-.007 (±.003)		.013* (±.005)	.020 (±.006)	.026* (±.007)
9	-.371** (±.026)	-.272** (±.026)	-.199** (±.026)	-.120** (±.020)	-.063** (±.013)	-.036** (±.008)	-.020* (±.005)	-.013 (±.005)		.006 (±.003)	.012 (±.004)
10	-.377** (±.026)	-.279** (±.026)	-.205** (±.026)	-.126** (±.021)	-.070** (±.014)	-.042** (±.009)	-.027* (±.006)	-.020 (±.006)	-.006 (±.003)		.006 (±.003)
11	-.383** (±.025)	-.285** (±.026)	-.211** (±.026)	-.132** (±.021)	-.076** (±.015)	-.048** (±.010)	-.033** (±.007)	-.026 (±.007)	-.012 (±.004)	-.006 (±.003)	

**P<.01 * P<.05

Pairwise comparison table depicting results based on the results of a two-way ANOVA depicting the relationship between Frequency Error and the nSweeps condition for the newborn data at one hundred iterations.

FE	1	2	3	4	5	6	7	8	9	10	11
1		-.362 (±.132)	-1.192** (±.179)	-2.456** (±.275)	-4.038** (±.327)	-5.008** (±.333)	-5.746** (±.350)	-6.111** (±.350)	-6.448** (±.372)	-6.783** (±.372)	-7.059** (±.408)
2	.362 (±.132)		-.830** (±.107)	-2.094** (±.212)	-3.676** (±.270)	-4.646** (±.294)	-5.384** (±.315)	-5.749** (±.322)	-6.086** (±.348)	-6.420** (±.357)	-6.697** (±.394)
3	1.192** (±.179)	.830** (±.107)		-1.264** (±.131)	-2.846** (±.190)	-3.816** (±.228)	-4.554** (±.269)	-4.919** (±.288)	-5.256** (±.322)	-5.591** (±.339)	-5.867** (±.380)
4	2.456** (±.275)	2.094** (±.212)	1.264** (±.131)		-1.582** (±.120)	-2.552** (±.180)	-3.290** (±.249)	-3.655** (±.279)	-3.992** (±.324)	-4.327** (±.352)	-4.603** (±.398)
5	4.038** (±.327)	3.676** (±.270)	2.846** (±.190)	1.582** (±.120)		-.970** (±.093)	-1.708** (±.166)	-2.073** (±.209)	-2.410** (±.253)	-2.745** (±.290)	-3.021** (±.335)
6	5.008** (±.333)	4.646** (±.294)	3.816** (±.228)	2.552** (±.180)	.970** (±.093)		-.738** (±.101)	-1.103** (±.144)	-1.440** (±.189)	-1.774** (±.229)	-2.051** (±.278)
7	5.746** (±.350)	5.384** (±.315)	4.554** (±.269)	3.290** (±.249)	1.708** (±.166)	.738** (±.101)		-.365** (±.075)	-.702** (±.105)	-1.036** (±.149)	-1.313** (±.192)
8	6.111** (±.350)	5.749** (±.322)	4.919** (±.288)	3.655** (±.279)	2.073** (±.209)	1.103** (±.144)	.365** (±.075)		-.337** (±.059)	-.671** (±.104)	-.948** (±.158)
9	6.448** (±.372)	6.086** (±.348)	5.256** (±.322)	3.992** (±.324)	2.410** (±.253)	1.440** (±.189)	.702** (±.105)	.337** (±.059)		-.334** (±.062)	-.611** (±.112)
10	6.783** (±.372)	6.420** (±.357)	5.591** (±.339)	4.327** (±.352)	2.745** (±.290)	1.774** (±.229)	1.036** (±.149)	.671** (±.104)	.334** (±.062)		-.276* (±.067)
11	7.059** (±.408)	6.697** (±.394)	5.867** (±.380)	4.603** (±.398)	3.021** (±.335)	2.051** (±.278)	1.313** (±.192)	.948** (±.158)	.611** (±.112)	.276* (±.067)	

**P<.01 * P<.05

Pairwise comparison table depicting results based on the results of a two-way ANOVA depicting the relationship between Slope Error and the nSweeps condition for the newborn data at one hundred iterations.

SE	1	2	3	4	5	6	7	8	9	10	11
1		12.303 (±4.422)	27.066** (±4.808)	49.608** (±)5.433	73.178** (±5.412)	86.089** (±5.832)	96.929** (±5.976)	102.529** (±6.098)	106.968** (±6.459)	110.740** (±6.751)	115.285** (±7.225)
2	-12.303 (±4.422)		14.762** (±2.527)	37.305** (±3.259)	60.875** (±3.573)	73.786** (±4.900)	84.626** (±5.043)	90.225** (±5.588)	94.665** (±6.078)	98.436** (±6.364)	102.982** (±6.915)
3	-27.066** (±4.808)	-14.762** (±2.527)		22.543** (±2.372)	46.112** (±2.992)	59.024** (±4.712)	69.864** (±4.978)	75.463** (±5.565)	79.902** (±6.237)	83.674** (±6.564)	88.219** (±7.052)
4	-49.608** (±5.433)	-37.305** (±3.259)	-22.543** (±2.372)		23.569** (±2.597)	36.481** (±3.655)	47.321** (±4.595)	52.920** (±5.254)	57.360** (±5.945)	61.131** (±6.394)	65.677** (±6.975)
5	-73.178** (±5.412)	-60.875** (±3.573)	-46.112** (±2.992)	-23.569** (±2.597)		12.911** (±2.860)	23.752** (±3.093)	29.351** (±3.896)	33.790** (±4.521)	37.562** (±5.052)	42.107** (±5.661)
6	-86.089** (±5.832)	-73.786** (±4.900)	-59.024** (±4.712)	-36.481** (±3.655)	-12.911** (±2.860)		10.840** (±2.169)	16.439** (±2.838)	20.879** (±3.402)	24.650** (±4.110)	29.196** (±4.833)
7	-96.929** (±5.976)	-84.626** (±5.043)	-69.864** (±4.978)	-47.321** (±4.595)	-23.752** (±3.093)	-10.840** (±2.169)		5.599* (±1.375)	10.039** (±1.810)	13.810** (±2.525)	18.356** (±3.202)
8	-102.529** (±6.098)	-90.225** (±5.588)	-75.463** (±5.565)	-52.920** (±5.254)	-29.351** (±3.896)	-16.439** (±2.838)	-5.599* (±1.375)		4.439 (±1.315)	8.211* (±2.131)	12.757** (±2.892)
9	-106.968** (±6.459)	-94.665** (±6.078)	-79.902** (±6.237)	-57.360** (±5.945)	-33.790** (±4.521)	-20.879** (±3.402)	-10.039** (±1.810)	-4.439 (±1.315)		3.772 (±1.332)	8.317* (±1.968)
10	-110.740** (±6.751)	-98.436** (±6.364)	-83.674** (±6.564)	-61.131** (±6.394)	-37.562** (±5.052)	-24.650** (±4.110)	-13.810** (±2.525)	-8.211* (±2.131)	-3.772 (±1.332)		4.546* (±1.109)
11	-115.285** (±7.225)	-102.982** (±6.915)	-88.219** (±7.052)	-65.677** (±6.975)	-42.107** (±5.661)	-29.196** (±4.833)	-18.356** (±3.202)	-12.757** (±2.892)	-8.31* (±1.968)	-4.546* (±1.109)	

**P<.01 * P<.05

Pairwise comparison table depicting results based on the results of a two-way ANOVA depicting the relationship between Tracking Accuracy and the nSweeps condition for the newborn data at one hundred iterations.

TA	1	2	3	4	5	6	7	8	9	10	11
1		.023 (±.008)	.064** (±.012)	.128** (±.017)	.213** (±.020)	.270** (±.020)	.318** (±.020)	.341** (±.021)	.365** (±.021)	.386** (±.022)	.408** (±.023)
2	-.023 (±.008)		.041** (±.006)	.105** (±.011)	.190** (±.014)	.247** (±.016)	.295** (±.017)	.318** (±.019)	.342** (±.019)	.363** (±.021)	.385** (±.023)
3	-.064** (±).012	-.041** (±.006)		.064** (±.008)	.148** (±.011)	.206** (±.014)	.253** (±.015)	.277** (±.018)	.301** (±.019)	.322** (±.021)	.344** (±.023)
4	-.128** (±.017)	-.105** (±.011)	-.064** (±.008)		.084** (±.007)	.142** (±.011)	.189** (±.014)	.213** (±.017)	.237** (±.019)	.258** (±.021)	.280** (±.024)
5	-.213** (±.020)	-.190** (±.014)	-.148** (±.011)	-.084** (±.007)		.057** (±.008)	.105** (±.011)	.128** (±.014)	.152** (±.016)	.174** (±.019)	.195** (±.021)
6	-.270** (±.020)	-.247** (±.016)	-.206** (±.014)	-.142** (±.011)	-.057** (±.008)		.048** (±.007)	.023** (±.010)	.095** (±.012)	.116** (±.015)	.138** (±.018)
7	-.318** (±.020)	-.295** (±.017)	-.253** (±.015)	-.189** (±.014)	-.105** (±.011)	-.048** (±.007)		.071* (±.006)	.047** (±.007)	.069** (±.010)	.090** (±.013)
8	-.341** (±.021)	-.318** (±.019)	-.277** (±.018)	-.213** (±.017)	-.128** (±.014)	-.071** (±.010)	-.023* (±.006)		.024** (±.004)	.045** (±.007)	.067** (±.011)
9	-.365** (±.021)	-.342** (±.019)	-.301** (±.019)	-.237** (±.019)	-.152** (±.016)	-.095** (±.012)	-.047** (±.007)	-.024** (±.004)		.022** (±.005)	.043** (±.008)
10	-.386** (±.022)	-.363** (±.021)	-.322** (±.021)	-.258** (±.021)	-.174** (±.019)	-.116** (±.015)	-.069** (±.010)	-.045** (±.007)	-.022** (±.005)		.021* (±.005)
11	-.408** (±.023)	-.385** (±.023)	-.344** (±.023)	-.280** (±.024)	-.195** (±.021)	-.138** (±.018)	-.090** (±.013)	-.067** (±.011)	-.043** (±.008)	-.021** (±.005)	

**P<.01 * P<.05

Appendix G

Exponential Modeling Tables

Tabular depiction of the exponential modeling outcomes for the adult ORI and NMF conditions at 1, 5, 10 and 100 Iterations.

	1 Iteration			5 Iterations			10 Iterations			100 Iterations		
	Tau	A_{noise}	A_{AS}	Tau	A_{noise}	A_{AS}	Tau	A_{noise}	A_{AS}	Tau	A_{noise}	A_{AS}
ORI FE	964	12.635	3.681	863	13.5	3.847	936	12.008	3.891	925	12.983	3.895
NMF FE	1071	6.693	3.644	463	8.356	3.585	648	8.067	3.67	700	7.743	3.721
ORI SE	521	-213.466	-49.15	472	-221.48	-53.115	488	-208.657	-50.722	760	-182.605	-53.037
NMF SE	1107	-77.357	-43.41	256	-104.963	-47.048	417	-103.203	-47.028	440	-99.873	-48.847
ORI TA	824	0.425	0.887	808	0.381	0.876	896	0.406	0.87	918	0.414	0.868
NMF TA	457	0.712	0.883	350	0.651	0.896	427	0.652	0.89	444	0.661	0.881

Tabular depiction of the exponential modeling outcomes for the adult hidden features at 1, 5, 10 and 100 Iterations.

	1 Iteration			5 Iterations			10 Iterations			100 Iterations		
	Tau	A_{noise}	A_{AS}	Tau	A_{noise}	A_{AS}	Tau	A_{noise}	A_{AS}	Tau	A_{noise}	A_{AS}
hFE	972	-5.634	-0.015	1153	-5.666	-0.064	1352	-5.147	-0.057	1194	-5.368	-0.073
hSE	577	117.31	5.343	795	102.178	4.224	742	95.312	4.849	823	96.525	3.554
hTA	751	0.012	0.34	1106	0.012	0.317	1577	0.009	0.269	1212	0.007	0.289

Tabular depiction of the exponential modeling outcomes for the newborn ORI and NMF conditions at 1, 5, 10 and 100 Iterations.

	1 Iteration			5 Iterations			10 Iterations			100 Iterations		
	Tau	A_{noise}	A_{AS}	Tau	A_{noise}	A_{AS}	Tau	A_{noise}	A_{AS}	Tau	A_{noise}	A_{AS}
ORI FE	2327	14.22	4.25	2939	13.76	4.16	2577	13.79	4.35	2722	12.49	4.29
NMF FE	2050	7.78	4.10	2241	8.68	4.03	1839	8.57	4.17	1598	8.84	4.28
ORI SE	634	-265.11	-74.60	2799	-216.34	-56.15	1958	-227.97	-62.06	1909	-222.65	-65.55
NMF SE	282	-142.20	-65.43	3446	-116.32	-55.46	2315	-117.38	-57.65	1636	-115.05	-60.15
ORI TA	2300	0.33	0.86	3221	0.333	0.8	3661	0.34	0.86	3636	0.35	0.85
NMF TA	2047	0.63	0.87	2215	0.55	0.87	2046	0.57	0.86	1662	0.57	0.85

Tabular depiction of the exponential modeling outcomes for the adult and newborn NMF condition at 1, 5, 10 and 100 Iterations.

	1 Iteration			5 Iterations			10 Iterations			100 Iterations		
	Tau	A_{noise}	A_{AS}	Tau	A_{noise}	A_{AS}	Tau	A_{noise}	A_{AS}	Tau	A_{noise}	A_{AS}
Adult FE	1071	6.69	3.64	463	8.35	3.58	648	8.06	3.67	700	7.74	3.72
Newborn FE	2050	7.78	4.10	2241	8.68	4.03	1839	8.57	4.17	1598	8.84	4.28
Adult SE	1107	-77.35	-43.41	256	-104.96	-47.04	417	-103.20	-47.02	440	-99.87	48.84
Newborn SE	282	-142.20	-65.43	3446	-116.32	-55.46	2315	-117.38	-57.65	1636	-115.05	-60.15
Adult TA	457	0.71	0.88	350	0.65	0.89	427	0.65	0.89	444	0.66	0.88
Newborn TA	2047	0.63	0.87	2215	0.55	0.87	2046	0.57	0.86	1662	0.57	0.85

Tabular depiction of the exponential modeling outcomes for the newborn hidden features at 1, 5, 10 and 100 Iterations.

	1 Iteration			5 Iterations			10 Iterations			100 Iterations		
	Tau	A_{noise}	A_{AS}	Tau	A_{noise}	A_{AS}	Tau	A_{noise}	A_{AS}	Tau	A_{noise}	A_{AS}
hFE	2968	-5.65	-0.06	3783	-5.37	-0.12	3460	-5.31	-0.17	3588	-5.25	-0.16
hSE	2178	108.77	0.576	2055	105.36	3.41	1884	110.59	3.79	2078	107.73	4.51
hTA	3136	0.01	0.24	7645	0.01	0.23	10776	0.01	0.24	6445	0.01	0.25

Tabular depiction of the exponential modeling outcomes for the adult ORI and NMF conditions at 1, 5, 10 and 100 Iterations.

	1 Iteration			5 Iterations			10 Iterations			100 Iterations		
	Tau	A_{noise}	A_{AS}	Tau	A_{noise}	A_{AS}	Tau	A_{noise}	A_{AS}	Tau	A_{noise}	A_{AS}
Adult hFE	972	-5.63	-0.02	1153	-5.66	-0.06	1352	-5.14	-0.05	1194	-5.36	-0.07
Newborn hFE	2968	-5.65	-0.06	3783	-5.37	-0.12	3460	-5.31	-0.17	3588	-5.25	-0.16
Adult hSE	577	117.31	5.34	795	102.17	4.22	742	95.32	4.84	823	96.52	3.55
Newborn hSE	2178	108.77	0.576	2055	105.36	3.41	1884	110.59	3.79	2078	107.73	4.51
Adult hTA	751	0.01	0.34	1106	0.01	0.31	1577	0.01	0.26	1212	0.01	0.28
Newborn hTA	3136	0.01	0.24	7645	0.01	0.23	10776	0.01	0.24	6445	0.01	0.25